RECENT PROGRESS
in
MANY-BODY THEORIES

Series on Advances in Quantum Many-Body Theory

Edited by R. F. Bishop, C. E. Campell, J. W. Clark and S. Fantoni
(International Advisory Committee for the Series of International Conferences on Recent Progress in Many-Body Theories)

Published

Vol. 1: Proceedings of the Ninth International Conference on Recent Progress in Many-Body Theories
Edited by D. Neilson and R. F. Bishop

Vol. 3: Proceedings of the Tenth International Conference on Recent Progress in Many-Body Theories
Edited by R. F. Bishop, K. A. Gernoth, N. R. Walet and Y. Xian

Vol. 4: Microscopic Approaches to Quantum Liquids in Confined Geometries
E. Krotscheck and J. Navarro

Vol. 5: 150 Years of Quantum Many-Body Theory
A Festschrift in Honour of the 65th Birthdays of John W Clark, Alpo J Kallio, Manfred L Ristig and Sergio Rosati
Raymond F. Bishop, Klaus A. Gernoth and Niels R. Walet

Vol. 6: Proceedings of the Eleventh International Conference on Recent Progress in Many-Body Theories
Edited by Raymond F. Bishop, Tobias Brandes, Klaus A. Gernoth, Niels R. Walet and Yang Xian

Vol. 7: Introduction to Modern Methods of Quantum Many-Body Theory and Their Applications
Adelchi Fabrocini, Stefano Fantoni and Eckhard Krotscheck

Forthcoming

Vol. 2: Microscopic Approaches to the Structure of Light Nuclei
Edited by R. F. Bishop and N. R. Walet

Vol. 8: Pairing in Fermionic Systems: Basic Concepts and Modern Applications
Edited by S. Armen, M. Alford and J. W. Clark

Series on Advances in Quantum Many-Body Theory – Vol. 9

Proceedings of the 12th International Conference

RECENT PROGRESS *in* MANY-BODY THEORIES

Santa Fe, New Mexico 23-27 August 2004

Editors

Joseph A. Carlson

Gerardo Ortiz

Los Alamos National Laboratory, USA

World Scientific

NEW JERSEY • LONDON • SINGAPORE • BEIJING • SHANGHAI • HONG KONG • TAIPEI • CHENNAI

Published by

World Scientific Publishing Co. Pte. Ltd.
5 Toh Tuck Link, Singapore 596224
USA office: 27 Warren Street, Suite 401-402, Hackensack, NJ 07601
UK office: 57 Shelton Street, Covent Garden, London WC2H 9HE

British Library Cataloguing-in-Publication Data
A catalogue record for this book is available from the British Library.

Series on Advances in Quantum Many-Body Theory — Vol. 9
RECENT PROGRESS IN MANY-BODY THEORIES
Proceedings of the 12th International Conference

ISBN 981-256-957-X

Printed in Singapore by World Scientific Printers (S) Pte Ltd

PREFACE

The 12th International Conference on Recent Progress in Many-Body Theories was held in Santa Fe, New Mexico in August 2004. The Conference offered an important opportunity to recognize important achievements in the field of quantum many-body theory in condensed matter and nuclear physics. It also offered the opportunity to present significant new results in many aspects of many-body theory and to highlight entire new fields which are strengthening both the breadth and depth of this discipline. Among these fields we would like to mention the emergent new areas of quantum information and ultracold atom physics, fields which are rapidly growing and exploring the frontiers of quantum physics.

At this conference, the Eugene Feenberg Medal conference was awarded jointly to Spartak T. Belyaev and Lev P. Gorkov. The citation reads as follows:

The 2004 Eugene Feenberg Medal is awarded jointly to Spartak T. Belyaev and Lev P. Gor'kov for their pioneering work on superfluidity and superconductivity. They independently introduced the revolutionary concept of anomalous propagators, thereby changing our understanding of the physics of quantum systems. The applications span all facets of the many-body problem: Belyaev explored dilute Bose liquids and pairing in nuclear matter, and Gor'kov laid down a powerful formulation of superconductivity that goes far beyond the original BCS work. Through the years, that formulation has become the standard language, as productive today as it was forty-five years ago. The terms of the Eugene Feenberg Medal specify that under exceptional circumstances, the award may be shared for a single corpus of work. Although Belyaev and Gor'kov worked independently, the conceptual unity of their approach is obvious. Together, they have changed many body physics and it is appropriate that they be honored together.

We were fortunate to have both Lev P. Gor'kov and Spartak T. Belyaev present at the conference, and their presentations are included in this volume. We are grateful to D. Pines and V. Zelevinsky for introducing Lev P. Gor'kov and Spartak T. Belyaev, respectively, and reviewing the history and outstanding accomplishments in their careers. We also thank Phil W. Anderson for presenting the conference Keynote Speech, a thought-provoking talk on present-day tribulations in physics with a positive outlook on the long-term future of science.

The main conference program was split into eight sections: Strongly Correlated Systems, Phase Transitions, Quantum Fluids and Solids, New Frontiers, Nuclear Physics and QCD, Cold Atoms, Quantum Information, and Complex Systems. Several of these sections are nearly or entirely new, and the talks necessarily generated a great deal of excitement. In particular the fields of Quantum Information and the rapid experimental and theoretical advances in Cold Atom physics were very well received by the participants. We were fortunate to have outstanding scientists and speakers which were able to convey the excitement of these fields.

Some of the other sections have historically had a strong representation at the conference, including Quantum Phase Transitions, Fluids and Solids, and Nuclear Physics. In these areas also a sense of rapid progress was apparent, as illustrated by the contributions in this volume. In nuclear physics and QCD, for example, talks focused on possible new exotic states of matter at high Density, using nuclei as tests of the Standard Model, and the rapid progress being made in large-scale simulations of lattice QCD.

This book summarizes many of the excitating topics presented during the course of the meeting.

In the general area of *Strongly Correlated Systems and Phase Transitions* the following subjects are reviewed: Cardy discusses from a many-body viewpoint a complementary way of looking at quantum critical systems in 1+1 dimensions by mapping them to one-dimensional Brownian motion problems. This mathematical technique goes under the name of Schramm-Loewner evolution. Senthil provides an overview of exciting recent developments on the physics which is outside the framework of the Landau paradigm of phase transitions. This analysis questions the validity of the applicability of Landau's ideas to certain strongly correlated systems, in particular, at zero temperature quantum phase transitions. Bishop reviews the powerful coupled-cluster method, illustrating its applicability to spin-lattice systems. Most importantly, he concentrates on showing the success of this technique to describe the competition between magnetic orders in frustrated lattices, systems where other computational methods are at a more primitive stage. Hallberg describes latest advances in the density matrix renormalization group technique with emphasis on the computation of dynamical properties. In particular she shows how this approach can be used in conjunction with Dynamical Mean-Field Theory to solve the associated impurity problem in the infinite-dimensional Hubbard model. In a somewhat similar spirit Honerkamp reviews the functional renormalization group method concentrating on its application to the two-dimensional fermionic Hubbard model. He also briefly describes a novel renormalization approach to explore the phase diagram of weakly interacting fermion systems. Chaplin digs into a subject that is not free of controversies.... Finally, Normand closes the section with an study and demonstration that Aharonov-Bohm rings of correlated electrons can be used as an experimental probe to detect spin-charge separation.

The section on *Quantum Fluids and Solids* gravitates around Helium, para-Hydrogen, and charged Van der Waals quantum films. Using an equation-of-motion method Krotscheck develops a systematic scheme to study excitations in strongly interacting Fermi systems. His analysis includes pair-excitations and compares his results to recent experimental measurements of the dynamical density-density structure function in two-dimensional ^{3}He. Manousakis studies the effect of adding a long-range Coulomb repulsive interaction to an attractive short-range potential (with a repulsive hard-core) in a Bose quantum film. He is interested in the potential competing, homogeneous and inhomogeneous (e.g. stripes), phases of this Van der Waals charged fluid. Bulk para-Hydrogen has been speculated to display a superfluid phase, however, this prediction has not been confirmed experimentally. Part of the problem is that the strength of the potential between Hydrogen molecules favors crystallization. Several efforts concentrated on artificially stabilizing a liquid superfluid phase. Boninsegni presents a Path-Integral Monte Carlo study of two-dimensional para-Hydrogen embedded in a crystalline matrix of Alkali atoms to argue that this particular system does not exhibit a superfluid phase. Vitiello concludes the section with an investigation of the effects of interatomic potentials in the equation of state of ^{4}He.

The section on *Nuclear Physics and QCD* combines includes talks on the physics of nuclei and QCD. Nuclei are mesoscopic systems which exhibit unique behaviors characteristic of their finite size. Iachello discusses the theory and experimental probes of phase transitions in nuclei. Nuclei are also important as a probe of new physics, Engel discusses the importance of nuclear physics in extracting neutrino masses from neutrinoless double-beta decay, and in the search for CP violation in nuclei. Traditionally, studies of the lighter nuclei and nuclear matter have been based upon microscopic approaches based upon the nucleon-nucleon interaction. Schwenk introduces different approaches to incorporating these interactions calculations of nuclear structure. Zelevinsky, on the other hand, explores what aspects of nuclear physics can be obtained through random interactions in a picture of single-particle orbitals. QCD is also playing a more important role in nuclear physics in recent times, even at very low energies and temperatures. In this volume Chandrasekharan explores new approaches to strong-coupling QCD, and Reddy explores new phases of QCD at high baryon density and low-temperature, phases which may be important in extreme astrophysical environments.

Cold Atoms and Quantum Information are two relatively new topics in this series, of growing importance to the topic of Quantum Many-Body theory. Experimentally an exciting new development is the ability to cool atomic Fermi gases to very low temperatures. Shlyapnikov provides a valuable description of the different regimes accessible to the experiments, from the BCS regime of weakly-interacting fermions, to the strongly-interacting regime near unitarity, to the BEC regime of bound molecules. Also the addition of optical lattices allows an impressive array of physical regimes to be studied, recent developments and future possibilities are discussed by Rolston. The conference also had intriguing talks on quantum information and quantum algorithms. Generalized quantum entanglement is an important problem in quantum computation, its relation to quantum phase transitions is discussed in a paper by Somma and collaborators. Hruska talked about a quantum non-demolition measurement of a single spin using Josephson oscillations, and Presilla contributed a talk on a probabilistic approach to quantum ground states in lattice systems.

The last section on *Complex Systems* addresses issues on dirty quantum Hall ferromagnets, disordered dissipative quantum systems, spin glasses, and the use of statistical physics methods in combinatorial optimization problems. Chalkers starts the section with an analysis of the interplay between ferromagnetic exchange and disorder in quantum Hall ferromagnets, i.e. two-dimensional electron systems in a perpendicular magnetic field in the presence of random impurities. What makes these systems particularly interesting is their peculiar relation between spin configurations (skyrmions) and charge density. The field of disordered quantum spin systems coupled to dissipative environments is the subject reviewed by Cugliandolo. She discusses several interesting out-of-equilibrium spin problems including the effect of quantum fluctuations on the real-time dynamics of glasses, and the fate of the Griffiths phase when the quantum spin system is coupled to an Ohmic bath. After critically reviewing the static properties of classical spin glasses Nishimori presents a conjecture on the exact location of the multicritical point in the phase diagram of finite-dimensional classical glasses. The theoretical framework he uses exploits duality and gauge symmetry and, if his conjecture proves indeed correct, it constitutes the first analytic result revealing the value of the critical point in finite-dimensional systems. The clever use of all the important methodological advances in the field of spin glasses in other arenas, such as combinatorial optimization problems, is the topic Zecchina sharply describes in the last contribution to this section. Using the MAX-K-SAT problem as an example he has shown that the Survey Propagation algorithm beats performance-wise other local-search algorithms.

We would like to thank all the speakers and other participants for making this conference so lively and exciting. We feel that the works presented in this volume faithfully represent the quality and breadth of the meeting. We also need to thank many who gave us help in organizing the program of this conference, and the International Advisory Committee for their support of helpful advice. Finally, we would like to thank Los Alamos National Laboratory, in particular the Theoretical Division, the Center for Non-Linear Science, and the Atomic, Nuclear, and Condensed Matter Groups in the division for their generous support which made this conference possible.

Joseph Carlson and Gerardo Ortiz

Editors

CONTENTS

Introduction

Preface v

International Advisory Committee xii

Feenberg Medal Session **1**

Surface and Superconductivity 3
L. P. Gor'kov

Spartak T. Belyaev — Recipient of the Feenberg Medal 8
V. Zelevinsky

Many-Body Physics and Spontaneous Symmetry Breaking 13
S. T. Belyaev

Keynote Speaker **25**

The Future Lies Ahead 27
P. W. Anderson

Strongly Correlated Systems and Phase Transitions **31**

Exact Results for Many-Body Problems Using Few-Body Methods 33
J. Cardy

Quantum Matters: Physics Beyond Landau's Paradigms 41
T. Senthil

Microscopic Calculations of Quantum Phase Transitions in Frustrated Magnetic Lattices 50
R. F. Bishop & S. E. Krüger

Recent Applications of the DMRG Method 62
K. Hallberg

Functional Renormalization Group in the 2D Hubbard Model 74
C. Honerkamp

Quantum Phase Transitions and Event Horizons: Condensed Matter Analogies 85
G. Chapline

Spin-Charge Separation and Topological Phase Transitions in Aharnov–Bohm Rings of Interacting Electrons 89
B. Normand, K. Hallberg, A. A. Aligia & A. P. Kampf

Quantum Fluids and Solids **95**

Two-Particle-Two-Hole Excitations in ^{3}He 97
E. Krotscheck, H. M. Böhm & K. Schörkhuber

Monolayer Charged Quantum Films: A Quantum Simulation Study 107
K. Wierschem & E. Manousakis

Can Inconmensuration Stabilize a Superfluid Phase of Para-Hydrogen? 117
M. Boninsegni

Analysis of the Interatomic Potential of the Helium Systems 122
S. Ujevic & S. A. Vitiello

Nuclear Physics and QCD **127**

Quantum Phase Transitions in Mesoscopic Systems 129
F. Iachello

Nuclear-Structure Theory in the Search for New Fundamental Physics 137
J. Engel

Matter at Extreme Density and its Role in Neutron Stars and Supernova 146
S. Reddy

New Approaches to Strong Coupling Lattice QCD 156
S. Chandrasekharan

Nuclear Interactions from the Renormalization Group 166
A. Schwenk

Random Interactions and Ground State Spin of Finite Fermi Systems 172
V. Zelevinsky & A. Volya

Cold Atoms and Quantum Information **181**

Superfluid Regimes in Degenerate Atomic Fermi Gases 183
G. V. Shlyapnikov

Bosons in Optical Lattices 199
S. L. Rolston

Generalized Entanglement and Quantum Phase Transitions 204
R. Somma, H. Barnum, E. Knill, G. Ortiz & L. Viola

Ground State of Many-Body Lattice Systems via a Central Limit Theorem 214
C. Presilla & M. Ostilli

Effects of a Single Quantum Spin on Josephson Oscillations 223
M. Hruška, L. Bulaevskii, A. Shnirman & D. Smith

Complex Systems 229

Spin Textures and Random Fields in Dirty Quantum Hall Ferromagnets 231
J. T. Chalker

Dissipative Quantum Disordered Models 241
L. F. Cugliandolo

Possibly Exact Solution for the Multicritical Point of Finite-Dimensional Spin Glasses 251
H. Nishimori, K. Takeda & T. Sasamoto

From Statistical Physics Methods to Algorithms 260
D. Battaglia, M. Kolař & R. Zecchina

INTERNATIONAL ADVISORY COMMITTEE

Participants of Many-Body XII; Santa Fe, NM; Aug 2004

Feenberg Medal Recipients, Keynote Speaker, and Organizers; MBXII; Santa Fe, NM.
(l to r) Gerardo Ortiz, Lev P. Gor'kov, P. W. Anderson, (statue), Spartak T. Belyaev, Joseph Carlson, & Eddy Timmermans

Feenberg Medal Session

SURFACE AND SUPERCONDUCTIVITY

L. P. GOR'KOV*

National High Magnetic Field laboratory,
Florida State University, Tallahassee, Florida

Received 1 November 2004

Experiments reveal the existence of metallic bands at surfaces of metals and insulators. The bands can be doped externally. We review properties of surface superconductivity that may set up in such bands at low temperatures and various means of superconductivity defection. The fundamental difference as compared to the ordinary superconductivity in metals, besides its two-dimensionality lies in the absence of the center of space inversion. This results in mixing between the singlet and triplet channels of the Cooper pairing.

Keywords: Superconductivity; surfaces; pairing.

1. Introduction

The purpose of a few remarks suggested below is to draw attention to phenomena that may take place at some metallic surfaces. By the latter I do not mean just surface effects on a bulk metal. There are currently numerous spectroscopic evidences (such, for instance, as APRES data, see, e.g. in 1), or STM results[2] indicating that so-called "dangling bonds" at a sharp surface, the concept familiar from the physics of semiconductors, may realize themselves at a surfaces of metals as well. These surface states form coherent metallic bands that for specific crystallographic orientations are well localized at the surface and do not overlap in energy with metallic bands inside the bulk.

Enormous literature is devoted to study of surface phenomena and its applications, such as the catalysis and the like. Below we will only discuss those few aspects that have common grounds with more general condensed matter physics. Specifically, we consider possibility of superconductivity (SC) in such localized bands on the surface of a metal (and even of an insulator[3]). We focus on some peculiarities by which the surface SC differs from SC in the bulk, and discuss means of its detection.

SC at a surface is a two-dimensional (2D) effect, and, strictly speaking, is not of a mean field type at finite temperature,[4] in other words, it has no exponential phase long-range order. Onset of a 2D SC occurs as the Berezinskii-Kosterlitz-Thouless transition. Nevertheless, one can analyze the 2D SC state at the surface by standard

*Also at L. D. Landau Institute for Theoretical Physics, 142432 Chernogolovka, Russia.

methods, even using the mean field (BCS-like) approach (at $T_c \ll \epsilon_F$). The reason for this is that loss of the phase coherence along the sample would take place for so long a distance that one may forget about the rigorous mathematical restriction of Ref. 4.

Examples of 2D quasi-two-dimensional (Q2D) superconductivity are now available in abundance. Organic materials, such as the Bechgaard salts $(TMTSF)_2PF_6$ with its $T_c \sim 1K$ under pressure $\sim$ 9kbar[5] and others, form the class of Q1D superconductors. Q2D charge transfer salts ($\kappa - (ET)_2I_3$ etc.) with $T_c \sim 8 - 12K$ provide another example.[6] High T_c cuprates with the conduction CuO_2 planes are nowadays the most common example of layered materials. Another 2D (layered) material, for instance, is Li-doped β-HfNCl with $T_c \approx 25K$.[7]

In most of the above examples a weak coupling between layers would result in a three-dimensional superconductivity at low temperatures. Surfaces (and superconductivity in metallic layer at the surface) are singled out and present a specific class, in particular due to their inherent "top-bottom" asymmetry. The lack of the inversion symmetry (CI-center of inversion) and its consequences for SC keep attracting interest of a number of groups. In what follows we briefly summarize some of the results.[8,9,10,11]

As it was shown 12, the order parameter of superconducting phase is the wave function of the Cooper pair:

$$\left\langle \hat{\Psi}_\alpha(\vec{r})\hat{\Psi}_\beta(\vec{r}\,')\right\rangle, \tag{1}$$

or in the momentum space (in the homogeneous state):

$$\left\langle \hat{\Psi}_\alpha(\vec{p})\hat{\Psi}_\beta(-\vec{p})\right\rangle \equiv -\left\langle \hat{\Psi}_\beta(-\vec{p})\hat{\Psi}_\alpha(\vec{p})\right\rangle, \tag{2}$$

where $\hat{\Psi}_\alpha(\vec{r})$ ($\hat{\Psi}_\alpha(\vec{p})$) are the electron field operators. Making use of the time reversal symmetry ($t \to -t$) and the space inversion (CI), it is easy to prove that Eq. (2) can be factorized as a product of the coordinate and spin components:

$$\left\langle \hat{\Psi}_\alpha(\vec{p})\hat{\Psi}_\beta(-\vec{p})\right\rangle = f(\vec{p})(i\hat{\sigma}_y)_{\alpha\beta}, \quad \text{if } S = 0, \tag{3}$$

and

$$\left\langle \hat{\Psi}_\alpha(\vec{p})\hat{\Psi}_\beta(-\vec{p})\right\rangle = \{(\vec{d}(\vec{p})\hat{\vec{\sigma}})i\hat{\sigma}_y\}_{\alpha\beta}, \quad \text{if } S = 1, \tag{4}$$

where $\hat{\sigma}$ are the Pauli matrices. The classification preserves itself even in the presence of spin-orbital interactions; the "singlet" ($S = 0$) and "triplet" ($S = 1$) pairings correspond then to the parity indices.

Most of known superconductors possess the center of inversion. An exception is $CePt_3Si$ ($T_c \approx 0.75K$).[13] On the other hand, at the surface the CI symmetry is always broken, and the new term (known as "Rashba term"[14]) can be added to the electron Hamiltonian:

$$\hat{H}_{SO} = \alpha(\hat{\vec{\sigma}} \times \vec{p}) \cdot \vec{n} \tag{5}$$

(here $\vec{p}$ lies in the plane, $\vec{n}$-the normal-to-plane vector).

The Kramers Theorem ($s = 1/2$) gives that at time inversion

$$(t \to -t): \; \hat{\Psi}_{\vec{p}} \to (\hat{\Psi}_{\vec{p}})^* = \hat{\Psi}^*_{-\vec{p}} \tag{6}$$

and at space inversion

$$\mathrm{CI}\ (\vec{p} \to -\vec{p}): \; \hat{\Psi}_{\vec{p}} \to \hat{\Psi}_{-\vec{p}} \neq \hat{\Psi}^*_{-\vec{p}}.$$

In other words, in the presence of both the time invariance and CI the electronic bands are double degenerate ($\vec{p}; \uparrow, \downarrow$). New term (5) lifts this degeneracy. For free electrons diagonalization of $H = H_0 + H_{SO}$ gives for the eigenfunctions $\eta_\lambda(\vec{p})$ and the eigenvalues $\epsilon_\lambda(p)$:

$$\eta_\lambda(\vec{p}) = \frac{1}{\sqrt{2}} \begin{pmatrix} 1 \\ i\lambda \exp(i\phi_{\vec{p}}) \end{pmatrix}$$
$$\epsilon_\lambda(p) = \epsilon_0(p) + \epsilon^{SO}_\lambda(p); \; \epsilon^{SO}_\lambda(p) = -\alpha\lambda p \tag{8}$$

($\lambda = \pm 1$).

We will follow Ref. 10 below to analyze how the lifting of the bands' degeneracy and the absence of CI caused by the new term (5) would mix together singlet and triplet pairings in (3) and (4):

$$\left\langle \hat{\Psi}_\alpha(\vec{p})\hat{\Psi}_\beta(-\vec{p}) \right\rangle \Rightarrow \left\{ if(\vec{p})\hat{\sigma}_y + (\vec{d}(\vec{p})\hat{\vec{\sigma}})i\hat{\sigma}_y \right\}_{\alpha\beta}. \tag{9}$$

Before to proceed, it is worth mentioning that the splitting, $2|\alpha|p_F$, may be rather significant, especially, for heavier elements. According to 1, it equals 0.11 eV for gold and reaches even larger values for Mo (0.13 eV) and W (0.5 eV). Therefore, the Fermi surfaces in (8) are quite independent, at least, from the viewpoint of superconductivity at low enough T_c.

We employ the weak coupling scheme:

$$\hat{H_{int}} = \frac{1}{2} \sum_{\lambda\mu\nu\rho} \sum_{\vec{p},\vec{p}',\vec{q}} U_{\lambda\mu\nu\rho}(\vec{p},\vec{p}',\vec{q}) \hat{a}^+_\lambda(\vec{p}) \hat{a}^+_\mu(-\vec{p}-\vec{q}) \hat{a}_\nu(-\vec{p}'-\vec{q}) \hat{a}_\rho(\vec{p}'), \tag{10}$$

where the first sum runs over all bands' indices. $U_{\lambda\mu\nu\rho}$ are the matrix elements of an interaction in terms of the eigenfunctions (8). For analysis of low T_c superconductivity we can neglect the non-diagonal parameters and take $\vec{q} = 0$ in (10) to simplify the Hamiltonian:

$$U \to U_{\lambda\mu\nu\rho}(\vec{p},\vec{p}') \equiv \frac{1}{2}\left[U_{\lambda\lambda\mu\mu}(\vec{p},\vec{p}') - U_{\lambda\lambda\mu\mu}(\vec{p},-\vec{p}')\right] \approx$$
$$\approx \frac{1}{2}\lambda\mu U(0) \exp\left[i(\phi_{\vec{p}'} - \phi_{\vec{p}}\right]. \tag{11}$$

In (11) we neglect all dependencies on momenta in the interaction leaving only its isotropic part, $U(0)$.

The set of Green functions, $g_\lambda(\vec{p},\omega_n)$, $f_\lambda(\vec{p},\omega_n)$, $f^+_\lambda(\vec{p},\omega_n)$ can be defined for each FS. The equations for them follow in the same fashion as in 12. After tedious but straightforward transformations it turns out that the energy spectrum (i.e. the thermodynamics!) and the self-consistency equations acquire the familiar form:

$$E_\lambda(\vec{p}) = \pm[v_F(p - p_\lambda)^2 + |\Delta(T)|^2]^{1/2}$$
$$-\frac{1}{2}TU(0)\sum_{n,\lambda,\vec{p}} \frac{1}{\omega_n^2 + \xi_\lambda^2 + |\Delta(T)|^2} = 1. \tag{12}$$

However, one obtains that, for instance, $F(\vec{p}, \omega)$ consists of two parts:

$$F_S(\vec{p}, \omega_n) = f_S(\vec{p}, \omega_n)\sigma_y; F_{AS}(\vec{p}, \omega_n) = -f_{AS}(\vec{p}, \omega_n)\sigma_y(\vec{\sigma} \cdot \vec{p}/|p|). \tag{13}$$

Functions f_S and f_{AS} are defined through the values f_λ for each band λ:

$$f_\lambda = f_S + \frac{1}{2}\lambda f_{AS} \tag{14}$$

Similar expressions follow for the remaining Green functions.

The part we would like to emphasize here is that both $G_{AS}(\vec{p}, \omega_n)$ and $F_{AS}(\vec{p}, \omega_n)$ have no smallness in α, i.e. they are of the order of unity, but only inside the narrow ring between two splitted Fermi surfaces in (8). They, however, change sign under transition from electrons to holes and, therefore, make no contribution into the gap value. Thus, the above analysis has shown that even for an interaction that would lead to s-wave (singlet) paring the H_{SO}-term results in admixing of a triplet contribution to the order parameter.

A few words about the origin of H_{SO}. Relativistic (Darwin) corrections to the electron Hamiltonian in vacuum are

$$\Delta\hat{H} = \frac{\hbar^2}{4m^2c^2}(\vec{\nabla}V \times \vec{p} \cdot \vec{\sigma}) \Rightarrow \frac{1}{2}\mu_B \vec{B} \cdot \vec{\sigma},$$

where $\vec{\nabla}V = -e\vec{E}$; $\vec{B} = \left[\vec{v} \times \vec{E}\right]/c$ are too weak to explain the effect.[1] Large value of α, however, comes about from the fact that the Darwin spin-orbit interaction becomes significantly enhanced inside the inner ionic orbitals in the solid.

Has the surface superconductivity ever been seen experimentally? We think that the answer should be positive: first, we refer to the experiments,3 where superconducting islands at the surface of the insulator WO_3 doped by Na ($Na : NO_3$) were seen by rather sophisticated STM experiments ($T_c \approx 91K$!). Second, superconductivity of compacted Pt powders has been observed in 15 at very low T_c ($\sim$1mK). SC for platinum has never been seen for the bulk. Superconductivity in Pt powder samples, on the other hand, may be easily explained as due to surface superconductivity: T_c depends on the packing fractions and increases for lower fractions.[15]

The above two examples demonstrate difficulties in observing superconductivity in atomic metallic layers at the surface. Indeed, the isolated surface metallic bands occupy a tiny layer of an atomic thickness. Traditional methods of detection such as measurements of anisotropy of the paramagnetic susceptibility or the thermodynamics of the surface SC phase seem to be very difficult because of the negligible contribution to be discerned on the background of the bulk properties. The Meissner screening would be difficult to see as well: imperfections of the surface itself most probably will result in formation of disconnected SC islands, although an incomplete screening could be observed.

There is, of course, the question of supercurrents and resistivity measurements. In the presence of the SC islands on the surface some percolation phenomena may be seen with temperature. In 11 the surface SC properties have been addressed theoretically in their dependence on a magnetic field applied along the surface. Without staying on the details, it was shown that in such a geometry new anisotropies appear as function of the magnetic field and depend on field orientation, akin to stripes caused by tendency to the LOFF inhomogeneity effects.

To conclude, possibility of superconductivity appearance in metallic 2D bands at surfaces was discussed. Occurrence of SC may be of importance for surface physics at very low temperature. The metallic bands can be doped by absorbents like cuprates. However, experimental study of SC in $Na : WO_3$ looks easier due to the fact that the bulk (WO_3-the tungsten bronze) is an insulator.

Acknowledgment

The work was supported by NHMFL through the NSF Cooperative agreement No. DMR-0084173 and the State of Florida.

References

1. S. La Shell et al., Phys. Rev. Lett. **77**, 3419 (1996); E. Rotenberg et al., Phys. Rev. Lett. **82**, 4066 (1999).
2. L. Petersen et al., Phys. Rev. B **57**, R6858 (1998).
3. S. Reich and Y. Tsabba, Eur. Phys. J. B **9**, 1 (1999); Y. Levi et al., Europhys. Lett. **51**, 564 (2000).
4. Hohenberg, Phys.Rev. **158**, 383 (1967).
5. e.g. see in the review: D. Jerome and H.J. Shulz, Adv. Phys. **31**, 299 (1982).
6. for review: T. Ishiguro, K. Yamaji, *Organic Superconductors*, volume **88** of the Springer-Verlag Series in Solid-State Science (Springer-Verlag Berlin Heidelberg, 1990).
7. S. Yamanako et al., Nature **321**, 580 (1998).
8. L.N. Bulaevskii et al., Sov. Phys. JETP **44**, 1276 (1976).
9. V.M. Edelstein, Sov. Phys. JETP **68**, 1244 (1989); Phys. Rev. Lett. **75**, 2004 (1995).
10. L.P. Gor'kov and E.I. Rashba, Phys. Rev. Lett. **87**, 037004 (2001).
11. V. Barzykin and L.P. Gor'kov, Phys. Rev. Lett. **89**, 227002 (2002).
12. L.P.Gor'kov, Sov. Phys. JETP **7**, 505 (1958).
13. E. Bauer et al., Phys. Rev. Lett. **92**, 027003 (2004).
14. Yu.A. Bychkov and E.I. Rashba, JETP Lett. **39**, 78 (1984).
15. R. Konig et al., Phys. Rev. Lett. **82**, 4528 (1999).
16. A. I. Larkin and Yu. N. Ovchinnikov, Sov. Phys. JETP 20, 762 (1965); P. Fulde and A. Ferrel, Phys. Rev. 135, A550 (1964).

SPARTAK T. BELYAEV — RECIPIENT OF THE FEENBERG MEDAL

VLADIMIR ZELEVINSKY

National Superconducting Cyclotron Laboratory and Department of Physics and Astronomy, Michigan State University, East Lansing, MI 48824-1321 USA

Received 1 December 2004

Ladies and gentlemen,

It is a great honor and privilege to take part in this session and to introduce the recipient of the Feenberg Medal, my teacher and friend, outstanding physicist Spartak BELYAEV. When ten years ago his colleagues from various countries got together for a conference in Philadelphia to celebrate his seventieth birthday, the name of the meeting was chosen to be "The Harmony of Physics". The word "Harmony" is the best characteristic of his life and his science, and his life is genuinely the life in science. Let me first draw the main lines of his biography and then briefly describe his principal scientific achievements.

Spartak Belyaev was born on October 27, 1923 in Moscow. He graduated from a high school on the eve of the war and already in August 1941 he entered the acting Army as a volunteer. He went through all life-threatening hardships of the battle, from sad retreats of 1941 in Caucasian mountains to the final offensive in Poland in 1945. Rejecting a promising career in the military, he entered the Physics Department of the Moscow University and already in 1947 started his research at the Laboratory #2 that later became Atomic Energy Institute, currently Russian Research Center "Kurchatov Institute".

Belyaev was lucky to begin his scientific life in the constellation of such bright (and very different) personalities as Gersh Budker, Arkady Migdal and Victor Galitsky. Those who survived the war matured much faster than the youngsters of today. Spartak's natural talent developed rapidly and successfully as if in compensation for the five lost years. Few declassified papers published in the proceedings of the Geneva conference contain amazing evidence of his great accomplishments. In 1958-1962, working in Moscow and, for a year, at the Niels Bohr Institute in Copenhagen, he made world-renowned breakthroughs in several directions of quantum many-body theory which brought him the general recognition at that time, Landau Gold Medal in 1998, and Feenberg medal in 2004.

In 1962 Spartak resolutely changed his life and moved to Siberia, where he was elected to the Russian Academy of Sciences, became Head of Theoretical Physics at the newborn Institute of Nuclear Physics (now Budker Institute) and later Rector of the very young Novosibirsk University. The Institute, lead by wisdom of Budker, with abundance of young gifted researchers, internal "disordered democracy", and location in the Academgorodok near Novosibirsk, far away from Moscow authorities, was at that time one of the best (I would dare to say "The best") place for science in the Soviet Union. The University was able to search for gifted youth everywhere to the East of Ural mountains, attract them, educate and involve into modern research. Still actively working in science, Spartak put a lot of efforts in creation and growth of the University. Now we can see Novosibirsk alumni as scientists of highest international standing spread all over the globe.

In my memory that time is indissoluble from the standard picture repeated daily: in the morning Spartak would enter my office and ask: "What's new?". Of course, I could answer "None", and nothing bad would happen. However, after few such responses, you start feeling uncomfortably and better try to have something new... And Spartak's reaction to anything really new was so immediate, and the started in this way discussions could continue into late night...

After 16 years in Siberia, Spartak returned to Moscow. And again his life is full with research, teaching and administrative duties. Until now he is Director of the Institute for Nuclear and General Physics (a big division of the Kurchatov Center), Rector of the Institute of Natural Science and Ecology, Chair of the department at the Moscow Physico-Technical Institute, Chair of the Council on Nuclear Physics etc. He chaired a special Committee for analyzing the Chernobyl accident and preventing its repetition. During hard years for Russian science, he applied his energy and common sense to preserve the Kurchatov center and to develop international cooperation (for example, he actively participates in the Heidelberg-Moscow experiment on double-beta decay and the search for the nature of neutrino.[25,29])

Spartak started his theoretical work in plasma physics and kinetic theory. Together with Budker, they for the first time rigorously derived the relativistic kinetic equation[1] and pointed out effective ways of the solution. A deep idea of treating multiquantum recombination in an ionized gas as a diffusion in energy space[2] was later widely used in quantum optics. The kinetics of plasma in strong magnetic fields was developed[3] with the use of a fruitful idea of considering the new quasiparticles, "larmorons", – magnetic circles instead of original electrons.

The two papers[5,6] by Belyaev on interacting Bose-gas, in parallel with the work by Migdal and Galitsky on Fermi-gas, laid in a foundation of the modern quantum many-body theory. The diagram techniques for the spontaneous symmetry breaking and corresponding anomalous propagators introduced by Belyaev, see also 9, became an important tool of the new theoretical arsenal of many-body physics and quantum field theory. Belyaev proved a theorem of survival for the Bose condensate even in the case of the strong interaction and found the lifetime of phonons, the so-called Belyaev damping. All this stuff is currently in the center of interest in relation to remarkable successes of experiments in atomic and molecular traps.

The fundamental Copenhagen paper[7] by Belyaev has put understanding of pairing correlations in nuclei on a firm basis. After qualitative ideas by Aage Bohr, Ben

Mottelson and David Pines on similarity of nuclear pairing to superconductivity of metals, Belyaev presented the full theory of the phenomenon and its manifestations in all nuclear properties – binding, excitation spectra, transition probabilities, collective modes, rotational features and so on. He also indicated finite size effects which distinguish the nuclear problem. This study, until now the most cited work in nuclear structure, immensely influenced the further progress of nuclear theory.

Latest works by Belyaev and his collaborators on structure 'of complex nuclei also contained outstanding and stimulating ideas: popular model "pairing plus quadrupole interaction"[8] that until now serves as a testing ground for more elaborate computations, bosonization of collective excitations and boson expansion techniques,[11,12,14,15,16] methods of deriving the collective Hamiltonian in many-body systems,[13,20] rotation as an intrinsic excitation[10,19] new mechanisms of nuclear reactions,[17] pairing in loosely bound nuclei[22] and so on. One of these papers[14] was selected to open the volume 1 of the new journal "Yadernaya Fizika". Among the works of the last decade I would mention the ideas of particle-like excitations in liquids,[24] application of neural networks to the signal recognition,[27] the scientific estimate of Chernobyl after-effects,[26] and the new theory of interaction of ultra-cold neutrons with condensed matter.[28] Spartak always considered physics as an experimental science. He actively participated in development of new experimental methods (source of polarized nuclei,[4] colliding beams and superthin nuclear target in a storage ring,[18] synchrotron rings[23] and so on). And today Spartak is full of ideas,[30] and, as usual, it is a great pleasure to suggest him a controversial topic and to see his heated interest and a stream of counter-suggestions.

It became a common place to blame the young generations for their disrespect and ingratitude to the elder, parents and teachers. The only consolation invented by the mankind is that the new generation, in turn, will take care of their children and in this way will pay its debt forward rather than backward. I am very happy to be able, on this superlative occasion, to pay at least a tiny fraction of my debt to my teacher in science and life, and congratulate him with this high and fully deserved award.

There is a Russian custom to accompany congratulations by flowers. Instead of live flowers, I would like to show a beautiful picture painted by a young lady whose father was a physics student, and later Associate Chair of the Physics Department at the same Novosibirsk University. The painter lives now in Tallahassee, Florida, being married to a theoretical physicist, who was a student and postdoc at Michigan State, and now, as junior faculty at Florida State, is working with myself further developing ideas started by Spartak Belyaev. And this is a genuine relay of generations.

Gold roses, 2004, oil on canvas, 24 × 30 in, by Natalia Andreeva.

Below I give a short selected bibliography in the chronological order.

References

1. S.T. Belyaev and G.I. Budker. Relativistic kinetic equation, *Sov. Phys. Doklady* **107**, 807 (1956).
2. S.T. Belyaev and G.I. Budker. Multiquantum recombination in an ionized gas, in *Plasma Physics and Problem of Controllable Thermonuclear Reactions* (Published by Academy of Sciences, Moscow, 1958) Vol. 3 p. 41.
3. S.T. Belyaev. Kinetics of an ionized gas in strong magnetic field, in *Plasma Physics and Problem of Controllable Thermonuclear Reactions* (Published by Academy of Sciences, Moscow, 1958) Vol. 3 p. 66.
4. B.P. Adyasevich, S.T. Belyaev, E.K. Zavoisky, and Yu.P. Polunin. Sources of polarized nuclei for accelerators, in *Nuclear Reactions at Low and Intermediate Energies* (Published by Academy of Sciences, Moscow, 1958) p. 87.
5. S.T. Belyaev. Application of the methods of quantum field theory to a system of bosons, *Sov. Phys. JETP* **34**, 417 (1958).
6. S.T. Belyaev. Energy-spectrum of a non-ideal Bose gas. *Sov. Phys. JETP* **34**, 433 (1958).
7. S.T. Belyaev. Effect of pairing correlation on nuclear properties, *Mat. Fys. Medd. Dan. Vid. Selsk.* **31**, No. 11 (1959).
8. S.T. Belyaev. On the nature of the first excited states in even-even spherical nuclei, *JETP* **39**, 1387 (1960).
9. S.T. Belyaev. Extension of diagram technique of field theory to Fermi systems with pairing phenomenon (with application to spherical nuclei), *Physica* **26**, 181 (1960).

10. S.T. Belyaev. Concerning the calculation of the nuclear moment of inertia, *Nucl. Phys.* **24**, 322 (1961).
11. S.T. Belyaev and V.G. Zelevinsky. Anharmonic effects of quadrupole oscillations of spherical nuclei. Nucl. Phys. **39** (1962) 582.
12. S.T. Belyaev. Some aspects of collective properties of nuclei, in *Selected topics in nuclear theory*, Inter. Atom. Energy Agency, Vienna (1963).
13. S.T. Belyaev. Time-dependent self-consistent field and collective nuclear Hamiltonian, *Nucl. Phys.* **64**, 17 (1965).
14. S.T. Belyaev. Nucleon-phonon Hamiltonian for spherical nuclei, *Sov. J. Nucl. Phys.* **1**, 3 (1965).
15. S.T. Belyaev and V.G. Zelevinsky. The Green's function method in a simple nuclear model, *Sov. J. Nucl. Phys.* **2**, 422 (1966).
16. S.T. Belyaev. *Collective excitations of nuclei* (Gordon and Breach, N.Y., 1966).
17. S.T. Belyaev. Effective two body forces in nuclei and gauge invariance, *Phys. Lett.* **28B**, 365 (1969).
18. S.T. Belyaev, G.I. Budker and S.G. Popov. The possibility of using storage rings with internal thin targets, in *High-Energy Physics and Nuclear Structure*(Plenum Press, N.Y., 1970) p. 606.
19. S.T. Belyaev and V.G. Zelevinsky. Rotation as intrinsic nuclear excitation, *Sov. J. Nucl. Phys.* **11**, 416 (1970).
20. S.T. Belyaev and V. G. Zelevinsky. Method of generalized density matrix in theory of collective excitations, *Sov. J. Nucl. Phys.* **16**, 657 (1973).
21. S.T. Belyaev and V.G. Zelevinsky. Niels Bohr and the physics of the atomic nucleus, *Sov. Phys. Uspekhi* **28**, 854 (1985).
22. S.T. Belyaev, A.V. Smirnov, S.V. Tolokonnikov, and S.A. Fayans. Pairing in atomic nuclei in the coordinate representation, *Sov. J. Nucl. Phys.* **45**, 783 (1987).
23. S.T. Belyaev, *et al.* Photonuclear experiments on the electron storage ring "Siberia-2", Kurchatov Institute of Atomic Energy, Moscow (1990).
24. S.T. Belyaev and Yu.K. Krasnov. Particle-like excitations in viscous liquids, *Sov. Phys. Doklady* **318**, 1362 (1991).
25. A. Balysh, M. Beck, S.T. Belyaev *et al.* The Heidelberg-Moscow double beta decay experiment with enriched 76Ge. First results, *Phys. Lett.* B **283**, 32 (1992).
26. S.T. Belyaev, V.F. Demin, and V.S. Osmachkin. Chernobyl accident: critical analysis of consequences and defence measures, *Atomic Energy* **83**, 393 (1997).
27. A.M. Bakalyarov, A.Ya. Balysh, S.T. Belyaev, and V.I. Lebedev. The use of artificial neuron nets for separating events by the shape of a signal, Kurchatov Institute of Atomic Energy, Moscow (1998).
28. A.L. Barabanov and S.T. Belyaev. Multiple scattering theory for slow neutrons (from thermal to ultracold), *Eur. Phys. J.* B **15**, 59 (2000).
29. H.V. Klapdor-Kleingrothaus,..., S.T. Belyaev *et al.* Latest results from the Heidelberg-Moscow double beta decay experiment, *Eur. Phys. J.* A **12**, 147 (2001).
30. S.T. Belyaev. Are quantum wave packets observable? Sokolov effect: puzzling generation of $2p$ states in the $2s$ hydrogen beam passing through a wide metal slit. *Eur. Phys. J.* D **25**, 247 (2003).

MANY-BODY PHYSICS AND SPONTANEOUS SYMMETRY BREAKING

SPARTAK T. BELYAEV

Russian Research Center Kurchatov Institute, Moscow 123182, Russia
bst@kiae.ru

Received 1 December 2004

The idea of spontaneous symmetry breaking in many-body physics from personal perspective (Bose-gas, nuclear structure and a new approach of Generalized Density Matrix).

Keywords: Symmetry breaking; Green's functions; density matrix; pairing; rotation.

1. Introduction

The notion of spontaneous symmetry breaking has quite general meaning, applicable both to physical phenomena and techniques used in theoretical physics. I start with short history of my personal experience with the problem.

My student years at Moscow State University (1946-1952) coincided with triumph of quantum electrodynamics. Curriculum of our new Physical-Technical Department required a part-time practice at a research institute. For me, it was the Institute nowadays named after Igor V. Kurchatov. After diffusion over some experimental laboratories, I found my place in the theoretical division. Our head, Arkady Migdal, was occupied by several problems, but for many years he had a favorite one, superconductivity. The new methods in QED, Green's functions, Dyson equations, Feynman graphs, were crying for adaptation to many-body problem. Naturally, Migdal with Victor Galitskii started such a work for Fermi systems. At first, my personal participation was only passive one, since I was deeply involved in plasma and beam physics working with Gersh Budker. Only after earning the doctorate (1955) I got free and decided to work on Bose system. At that time Migdal and Galitskii already found how to adjust the Green's function technique for many-body problems, having as their goal Fermi systems.[1] Very soon I realized that Bose systems are quite different.

2. Bose Gas

Peculiarity of a Bose system comes from a large, microscopic number of particles in the $\mathbf{p}=0$ state (*condensed phase*). This prevents direct use of Green's function technique. My idea was to split the system into two parts, the condensate, $\mathbf{p}=0$, and the rest, $\mathbf{p}\neq 0$, and consider them separately.

For the condensate one should introduce *n-body* Green's functions. Since they depend only on time variables and there is a small parameter, N^{-1}, one gets a hope. Accurate calculations turned out to be quite cumbersome. But the result was simple and physically transparent: to order $\sim N^{-1}$, the n-particle function has the form

$$iG_0(t_1...t_n;t'_1...t'_n) = K(t_1)...K(t_n)K^\dagger(t'_1)...K^\dagger(t'_n), \tag{1}$$

where

$$K(t) = \sqrt{n_0}\,\exp(-i\mu t), \tag{2}$$

and the chemical potential, $\mu = E_{N+1} - E_N \approx \partial E/\partial N$, can be calculated as $\mu = \sum(\text{loops})$, and the only new parameter is n_0, the density of the condensate.

The physics of eq. (1) can be understood from a particular case of a one-particle function in a non-interacting system, $iG_0^{(o)}(t-t') = \langle a_0(t)a_0^+(t')\rangle$, that describes the propagation of a particle from t' to t. In vacuum this process can proceed only by creating a particle at time t' and annihilating it at a later time t; in this case $G_0^{(o)}$ coincides with that in QED. But in the presence of N particles of the same type, the absorbed particle may be any one from the condensate. Time ordering of two events, creation at t' and absorption at t, is immaterial, within accuracy $\sim 1/N^{-1}$. That means the absence of correlation between creation and absorption processes, and the product form (1) corresponds to this physics. The use of (1) is equivalent to a simple ansatz: all operators a_0 and $a_0^\dagger$ should be replaced by

$$a_0(t) \Rightarrow \sqrt{n_0}\exp(-i\mu t), \quad a_0^\dagger(t) \Rightarrow \sqrt{n_0}\exp(i\mu t), \tag{3}$$

which is formally equivalent to a replacement of the quantum system, the condensate, by a uniform time-dependent field.

Exclusion of the condensate allows for direct application of QED techniques to the system of particles with $p \neq 0$, though with a new kind of elementary graphs and irreducible self-energy parts Σ_{ik} in Dyson-type equations. In addition to the ordinary Σ_{ik}, with an equal number of incoming and outgoing external lines ($i = k$), there are now self-energy parts Σ_{ik} where some external lines are replaced by the condensate "external field", and the numbers of incoming and outgoing lines may be different. For the single-particle Green's function, it is necessary to consider Σ_{11}, Σ_{20}, and Σ_{02}. The single-particle Green's function, disentangled from two coupled Dyson equations, was found as

$$G(p^0+\mu,\mathbf{p}) = \frac{p^0 + \varepsilon_\mathbf{p} + \Sigma_{11}^- - \mu}{\left[p^0 - (\Sigma_{11}^+ - \Sigma_{11}^-)/2\right]^2 - \left[\varepsilon_\mathbf{p} + (\Sigma_{11}^+ + \Sigma_{11}^-)/2 - \mu\right]^2 + \Sigma_{20}\Sigma_{02} + i\delta},$$

where $\Sigma_{11}^\pm = \Sigma_{11}(\pm p + \mu)$.

In the "gaseous" (low density) approximation, Σ_{ik} and μ were found as functions of the scattering amplitude $f(\mathbf{p}',\mathbf{p})$ and condensate density n_0. Quasiparticle excitation energy in the first order in $\sqrt{n_0 f_0^3}$ has the form ($m = 1$, $\hbar = 1$)

$$\epsilon(\mathbf{p}) = \sqrt{\left[\varepsilon_\mathbf{p} + 2n_0 f(\frac{\mathbf{p}}{2},\frac{\mathbf{p}}{2}) - n_0 f(00)\right]^2 - n_0^2\,|f(\mathbf{p},0)|^2}. \tag{4}$$

The second order approximation in $\sqrt{n_0 f_0^3}$ results in attenuation of quasiparticles (decay into two quasiparticles of lower energy). For small momentum it was found

$$\epsilon(\mathbf{p}) \approx p\sqrt{n_0 f_0}\left(1 + \frac{7}{6\pi^2}\sqrt{n_0 f_0^3}\right) - i\frac{3}{640\pi}\sqrt{n_0 f_0^3}\,\frac{p^5}{(n_0 f_0)^{3/2}}. \qquad (5)$$

The Green's function allows one to find other physical quantities of interest, density and chemical potential:

$$n - n_0 \approx \frac{n_0}{3\pi^2}\sqrt{n_0 f_0^3}, \quad \mu = n_0 f_0\left(1 + \frac{5}{3\pi^2}\sqrt{n_0 f_0^3}\right). \qquad (6)$$

Apart from the scattering amplitude, eqs. (5) and (6) contain only one parameter $n_0 f_0^3$, $f_0 = f(0,0)$. Just this parameter should be small for the validity of the gaseous approximation. It was not directly required that the total density n be small but eq. (6) shows that when n_0 is small, then n is also small. This means that it is not possible to reduce significantly the density of the condensed phase by increasing the interaction or the total density.

One general comment: solution of the problem becomes possible due to the replacement of the condensate, a quantum system, by a time-dependent external field due to the ansatz (3). Though quantitatively this operation is accurate up to the order $\sim N^{-1}$, it breaks the strong symmetry, total number conservation. Phonon-like excitations ($\epsilon(\mathbf{p}) \Rightarrow 0$ when $\mathbf{p} \Rightarrow 0$) emerge as a response to the symmetry breaking. This is a particular example of the general theorem on unavoidable appearance of such a mode (later on called *Goldstone mode*) after symmetry breaking.

One may look at the problem from a different point of view. The chemical potential μ is the only link between two subsystems (particles with $\mathbf{p} \neq 0$ and condensate). This is in full analogy with thermodynamic equilibrium of liquid with its vapor. Far from the phase transition, when fluctuations are small, properties of this vapor practically coincide with those of pure gas, but technically the gas is considered at fixed N, whereas the vapor at fixed μ.

Just as in the last example, one may from the beginning consider only the subsystem with $\mathbf{p} \neq 0$ at fixed μ, using the "corrected" Hamiltonian $\hat{H}' = \hat{H} - \mu\hat{N}'$ with $\hat{N}' = \sum a_{\mathbf{p}}^\dagger a_{\mathbf{p}}$, $\mathbf{p} \neq 0$, and operators a_0, $a_0^\dagger$ replaced by a c-number $\sqrt{n_0}$. The Hamiltonian $\hat{H}'$ allows for the use of the QED techniques. Two free parameters, n_0 and μ, can be then found using $\langle n' \rangle = n - n_0$ and minimization of ground state energy. Such an approach was demonstrated by Hugenholtz and Pines, Ref. 2.

Two papers with the results on Bose systems were sent for publication in summer 1957.[3,4] Before that, I gave a talk at the Landau weekly seminar after a preliminary personal discussion with him. (This was the usual procedure for all to-be-speakers at the Landau seminar). It was the time when the BCS theory of superconductivity became known in Moscow and excited the physical community. Naturally, I start to think on possible application of my results to superconductivity. But external circumstances made a different decision for myself. Director of the institute, I.V. Kurchatov, decided (and made it possible) to resume collaboration with European research centers. So, two young physicists, Vilen Strutinsky and myself, were sent for a year to the Niels Bohr institute in Copenhagen. (A few month later Victor Galitskii was sent to Rudolph Peierls in Birmingham).

3. 1957-58, Niels Bohr Institute

The main activity, research and discussions there were around the *unified nuclear model* by Aage Bohr and Ben Mottelson. The general scheme was clear and well confirmed. The base is a spherical mean field with shell structure and its evolution with deformation. It allows one to establish quantum numbers of odd nuclei and single-particle excitations for spherical and deformed nuclei. The main interest was for collective excitations, vibrations in spherical nuclei, and vibrations + rotation in deformed ones, since their properties, calculated with different residual interactions, did not agree with experiments. Among the puzzling theoretical results, in particular, were predictions for inertial parameters for vibrations and rotation.

After the BCS theory of superconductivity, Bohr, Mottelson and Pines made a suggestion[5] that the energy gap found in the spectra of even-even nuclei may be caused by correlation effects similar to those in superconductors. It was extensively discussed, with the use of the *seniority* quantum number, what is the analogue of a Cooper pair in the shell model (say, in a single j-level), how pairs occupy the shell etc. I decided to enter the problem from another, many-body, direction, thinking, as was later formulated in my article 6, that

... such correlations may also effect other nuclear properties, which have no analogue in superconductors, connected with the finite size of nuclei and the shell structure of the single particle levels.

In particular, I bore in mind long standing and puzzling theoretical results (obtained in fact for any reasonable interaction):

- Spherical shape is stable only for closed shell nuclei. Even one-two valence particles result in the deformation that gradually increases with growing occupation.
- The inertial parameter for vibrations is of hydrodynamic value (too small).
- The rotational moment of inertia is close to the rigid body value (too large).

Results of my attempts may be understood from *Concluding Remarks* in Ref. 6:
Starting from basic assumption that a pairing correlation of a "superconducting" type exists between nucleons, we have attempted to investigate consistently the effects of this correlation in different nuclear phenomena. Although the calculations are based on a rather idealized model, a great number of experimental facts of a different kind are explained in a natural way from a single point of view, viz.,

(a) Stability of the spherical shape of nuclei near the closed shells;
(b) Sharp transition between spherical and deformed nuclei;
(c) Significant reduction of the moment of inertia from the value for rigid rotation;
(d) Existence of low-energy vibrations in spherical nuclei near the bound of instability.

A few comments in addition. The pairing correlations were treated as in the BCS theory with particle number non-conservation (μ-*system* instead of N-*system*). The finite size of the nuclear mean field breaks also translational symmetry; a deformed mean field breaks rotational symmetry. Collective low energy vibrations (inside the gap) and rotations are just the response to this symmetry breaking, that is Goldstone modes. The problem of rotation has its own peculiarity that until now stimulates new theoretical approaches. First, let us cast a glance on existing many-body techniques.

4. Microscopic Many-Body Theory: A General View

In practice, the starting point and the base is always a mean field (MF) and the view on the system as composed of independent quasiparticles (QP). The MF is supposed to be, at least in principle, self-consistently determined from the same QP. At the next stage one can explore symmetry and dynamic properties of the MF, say, by its response to an external field, and recognize the structure of collective excitations. An external field for exploring the response of the system may be rather specific. One example, considered above, is the condensate in Bose system, acting on the subsystem with $\mathbf{p} \neq 0$. The second example, rotation induced by an external torque (*cranking*), will be considered below.

The Hartree-Fock (HF) approach to the self-consistent calculation of MF can be formulated as an equation for the density matrix $\rho_{12} = \langle 0|a_1^\dagger a_2|0\rangle$,

$$[h(\rho), \rho] = 0; \quad \rho^2 = \rho; \quad h(\rho_a) = \varepsilon_a + \mathrm{Tr}_b\,(V_{ab}\rho_b), \tag{7}$$

where the mean-field Hamiltonian $h(\rho)$ contains kinetic energy ε and the self-consistent potential, generated by the interaction V_{ab}. The HF method can be formulated also as a variational problem with fixed N.

The Hartree-Fock-Bogoliubov (HFB) method is an extension of HF, which includes Cooper pairing, when the N-violating pair transfer matrix elements $\langle N-2|\Delta|N\rangle$ and $\langle N+2|\Delta^\dagger|N\rangle$ are large. In variational form one looks for $E_0(\mu) = \min\langle \hat{H} - \mu\hat{N}\rangle$ with μ as a Lagrange multiplier to be determined from $\langle \hat{N}\rangle = N$ or $\partial E_0/\partial\mu = -N$. In terms of statistical thermodynamics, the transition from HF to HFB means replacement of the system under investigation. The system with a fixed number of particles and Hamiltonian H is replaced by the system with a fixed chemical potential, μ, with the grand canonical potential $H - \mu N$. Naturally, the solution for the μ-system gives only average quantities over few neighboring N-systems. Usually such an average description is satisfactory, but for nuclei near the spherical-deformed transition and weakly bound nuclei the pairing should be treated more accurately.[7]

Rotation seems to allow the same approach as pairing. In this case there are large matrix elements of the quadrupole moment $\langle J|Q_{2\lambda}|J'\rangle$. Therefore it seems possible to look for $E(\Omega) = \min\langle \hat{H} - \Omega\hat{J}\rangle$, with a Lagrange multiplier, angular velocity Ω, to be found from $\langle \hat{J}\rangle = J$, or $\partial E_0/\partial\Omega = -J$. In spite of formal similarity to the pairing case, rotation is much more complicated, since the angular momentum is the vector with non-commuting components. Usually only one projection is used assuming plane rotation of a nucleus with axially symmetric deformation, which allows one to find one moment of inertia. This may be reasonable for even-even nuclei when adiabatic rotation does not excite intrinsic degrees of freedom. But nuclear rotation turns out to be a phenomenon much more diverse and rich.

5. Nuclear Rotation

During last two-three decades, a lot of experimental data were collected mostly with large γ-ray arrays. Hundreds of new collective bands were found and precisely measured up to very high angular momenta. Among the new discoveries are *superdeformation*[8] (with axis ratio $\sim 2:1$), *identical bands*[9] (which coincide in neighboring nuclei with accuracy 0.1%), *bifurcation*[10] (small splitting at some value of I of a band with spins $I = 2n$ into two sub-bands with $I = 4n$ and $I = 4n + 2$), *backbending*[11] (transition from a band with pure rotational I to a band based on a quasiparticle pair with high intrinsic angular momentum), *magnetic rotation*[12] (with strong $M1$ transition matrix elements), *wobbling motion*[13] etc.

How adequate are theoretical tools for the richness of experimental data? One should distinguish between phenomenological and microscopic approaches. From symmetry and specific features of a rotational band, one can guess on the structure of a *phenomenological Hamiltonian* for its description, e.g. of the form

$$H = AI^2 + BI^4 + C(I_+^2 + I_-^2) + D(I_+^4 + I_-^4), \tag{8}$$

where the coefficients are to be chosen to fit experimental data, say near a bifurcation point. This might be reasonable provided the coefficients are determined consistently from intrinsic nuclear structure. Otherwise, it is a pure artificial problem of a rotor with a specific Hamiltonian.

Particle-rotor model (PRM))[14] is an attempt to describe the interaction between the rotor and an odd particle in the form

$$H_{\text{PRM}} = \frac{1}{2\mathcal{J}}(\mathbf{I} - \mathbf{j})^2. \tag{9}$$

This model permits one to analyze interplay between competing alignment forces, deformation and Coriolis, acting on a particle. But the restriction to Coriolis, a pure kinematical interaction with an unidentified rotor, does not allow for using the PRM for quantitative description of rotational spectra in odd nuclei.

The *cranking model* (CM) is, in fact, the only practical tool for microscopic calculations of rotational phenomena. One starts with a many-body Hamiltonian with an external torque,

$$\hat{H}(\mathbf{\Omega}) = \hat{H} - \mathbf{\Omega}\hat{\mathbf{J}}, \tag{10}$$

and proceeds with HF or HFB. This results in the *shell model in the rotational frame*, where the MF and single-particle levels are functions of angular velocity $\mathbf{\Omega}$, that is supposed to be excluded from final results and replaced by the angular momentum. The problem is that such a replacement cannot be done uniquely. In the traditional CM[15] one assumes the rotational axis coincide with one of the principal axis of the deformed field (*principal axis cranking*, PAC). For complicated rotational bands, for example, in odd-odd nuclei, the direction of the rotational axis is a subject for adjustment to fit experimental spectra (*tilted axis cranking*, TAC)[16]. Defects of CM are evident: it violates rotational symmetry, does not provide a regular way to estimate and correct the errors and prevents direct application to the problems of interaction between rotational and intrinsic excitations. "Average over J" description, imminent in the CM, becomes too crude for the cases where an interplay between rotation and intrinsic excitations results in an abrupt change in the band structure, an analogue of a phase transition.

6. Interim Summary

- Sacrifice of strong symmetry opens way for simple theoretical description of spontaneous symmetry breaking. The symmetry is controlled *in average* by Lagrange multipliers, which have a definite physical meaning. This can be interpreted as a replacement of the system under investigation.
- The method is satisfactory for macroscopic systems, where an average statistical description is the conventional one.
- In mesoscopic finite systems the description "in average" gives a crude approximation, especially in cases with sharp structural changes.
- A striking example is given by diversity of nuclear rotation, where many rotational bands demonstrate sharp structural changes.
- Microscopic methods with symmetry conservation are evidently needed.

A basic microscopic approach was suggested by Kerman and Klein[17] and developed by Klein with collaborators.[18] The starting point is equations of motion (EoM) for the matrix elements of single-particle creation and annihilation operators with precise conservation of angular momentum and particle number. Unfortunately, even after many years of reformulation, the method is still too complicated to serve as a practical tool. A similar, *generalized density matrix* (GDM) approach, based on EoM for bilinear combinations $aa^\dagger$ was suggested by Zelevinsky and myself.[19] Solutions obtained for several problems[20,21,22] demonstrated its practical efficiency. Since the method is not familiar to wider theoretical community and seems to be underestimated, I try to explain main ideas of the GDM approach.

7. Generalized Density Matrix (GDM)

Systems with spontaneous symmetry breaking always have a set of states (Goldstone modes) $|\Lambda\rangle$ strongly linked by large matrix elements (as $\langle N-2|\Delta|N\rangle$ for pairing and $\langle J|Q_{2\lambda}|J'\rangle$ for rotation). Therefore, we need a method which can handle matrix elements $\langle\Lambda|a_1 a^\dagger_{1'}|\Lambda'\rangle$, diagonal and off-diagonal, on the same footing as expectation values for the ground state in the HF method. The first step in this direction is known as the random phase approximation (RPA), when, for small amplitude vibrations, we assume for off-diagonal matrix elements

$$\langle 0|a_1 a^\dagger_{1'}|\Lambda\rangle \ll \langle 0|a_1 a^\dagger_{1'}|0\rangle \approx \langle\Lambda|a_1 a^\dagger_{1'}|\Lambda\rangle).$$

It results in the equation for the *transition density matrix*, $\langle 0|a_1 a^\dagger_{1'}|\Lambda\rangle$.

The GDM offers a possibility to go beyond mean field and RPA. The basic assumption is the possibility to separate from the Hilbert space of the many-body system a restricted *collective band* of states strongly linked by dynamics among themselves but only weakly with the states outside the band. In other words, dynamics may be saturated inside the collective band; in this language the HF method means the saturation of the dynamics by the ground state. The outcome is the nonlinear equation of the form

$$[H + h(R),\, R] = 0, \tag{11}$$

where $h(R)$ has structure similar to (7). All quantities in (11) are matrices in *combined* space, the product of single-particle ("intrinsic"), $|1\rangle$, and collective, $|\Lambda\rangle$, spaces. To emphasize this fact, we introduce, for all quantities in combined space, a convenient notation

$$\langle\Lambda|R(11')|\Lambda'\rangle \Rightarrow \langle\Lambda 1|R|1'\Lambda'\rangle.$$

Here the GDM operator R is a bilinear combination of creation and annihilation operators. If pairing is of importance, it is useful to consider the GDM in the "spinor" representation,

$$\hat{R}(1,2) = \begin{pmatrix} a_1 a_2^+ & a_1 a_{\tilde{2}} \\ a_{\tilde{1}}^+ a_2^+ & a_{\tilde{1}}^+ a_{\tilde{2}} \end{pmatrix}, \tag{12}$$

where tilde means time-conjugation. The Hamiltonian $H \Rightarrow \langle\Lambda|\mathcal{H}|\Lambda'\rangle$ depends, naturally, only on collective quantum numbers Λ, and is reduced to a diagonal matrix when $|\Lambda\rangle$ are eigenvectors of $\mathcal{H}$.

Equation (11) is of the HF type but in combined space, where $H + h(R)$ plays the role of the self-consistent MF Hamiltonian with the single-particle part $h(R)$ and collective part H. Coupling between single-particle and collective degrees of freedom reveals in dependence of $h(R)$ not only on single-particle variables but also on collective ones. Like in ordinary HF theory, eq. (11), should be supplied with a normalization condition for the GDM, $R^2 = R$. The GDM $\langle\Lambda 1|R|1'\Lambda'\rangle$ has a simple physical meaning: the trace (qR) over intrinsic space with a one-body operator (q), say quadrupole moment q_{ik}, results in the operator of the same type in collective space (Q_{ik}).

7.1. *Nuclear rotation revisited*

Select, as an example, the collective band as an isolated rotational band of a well deformed nucleus. At the next stage we can use the symmetry of the band governed by the rotational group $SU(2)$. The states within the band are uniquely labelled by the total spin J and its projection J_z onto the laboratory-fixed axis z, dynamic variables are Euler angles. We outline several steps which lead to solution the problem.

(a) Angular momentum conservation in combined space can be written as

$$[\mathbf{J}+\mathbf{j},R]=0; \quad [\mathbf{J}+\mathbf{j},h(R)]=0. \tag{13}$$

(Simultaneous rotation in both intrinsic and collective spaces cannot change R).

(b) There exists a transformation D, unitary in the intrinsic space and acting as a "shift" in the collective space,

$$D^{-1}(J_i+j_i)D=J_i. \tag{14}$$

For the transformed quantities $D^{-1}RD=R'$ and $D^{-1}hD=h'$, one has

$$[\mathbf{J},R']=0; \quad [\mathbf{J},h']=0,$$

which means that R' and h' do not depend on Euler angles, and the only possible dependence on collective variables is that on I_α, angular momentum projections on body-fixed axes. The collective Hamiltonian, on the other hand, is a function of components I_α transformed as

$$D^{-1}I_\alpha D=I_\alpha-j_\alpha, \tag{15}$$

so that, after the D-transformation, the collective Hamiltonian H depends also on intrinsic angular momentum.

(c) Due to rotational invariance of ε and the two-body interaction, one has $D^{-1}\varepsilon D=\varepsilon$; $D_a^{-1}D_b^{-1}V_{ab}D_bD_a=V_{ab}$, so that

$$h'(R)=D^{-1}hD=h(R'), \tag{16}$$

and the GDM equation after the D-transformation takes the form

$$[H(I_\alpha-j_\alpha)+h(R'),\ R']=0. \tag{17}$$

7.2. *Solution of the equation*

We start with a selection of the operator form of the collective Hamiltonian H. For instance, with the help of theoretical vision and/or experimental data we assume H in the form of expansion

$$H(I)=H_0+\frac{1}{2}H_{ik}I_iI_k+\frac{1}{4}H_{iklm}I_iI_kI_lI_m+\ldots \tag{18}$$

with coefficients to be found.

(ii). We look for the solution of (17) as an expansion

$$R'=R_0+R_i(I_i-j_i)+R_{ik}(I_i-j_i)(I_k-j_k)+\ldots. \tag{19}$$

With the help of eq. (16), it is easy to see that $h(R')$ is expanded as

$$h(R')=h_0+h_i(I_i-j_i)+h_{ik}(I_i-j_i)(I_k-j_k)+\ldots \tag{20}$$

with the coefficients

$$h_0 = \varepsilon + \mathrm{Tr}\,(VR_0) \equiv \varepsilon + w(R_0); \quad h_i = w(R_i); \quad h_{ik} = w(R_{ik}); ... \tag{21}$$

The main equation (17) with the ansatz of (20) and (21) is split into a set of equations according to the powers in $(I_i - j_i)$. The scalar equation for R_0 is non-linear being of the HF type; the vector equation for R_i and higher tensor equations are linear equations of the RPA type with right hand sides found at the previous steps. Solutions obtained for R_i, R_{ik} ,... depend on the parameters H_{ik}, H_{iklm},., which should be established from a consistency condition.

7.3. *Consistency condition*

In the basic equation (11), the collective Hamiltonian was defined as a matrix $H = \langle\Lambda|\mathcal{H}|\Lambda'\rangle$, that is, as a "projection", mapping of the many-body microscopic Hamiltonian $\mathcal{H}$ onto the collective band. In terms of the density matrix R, this condition can be written as

$$H \Leftrightarrow \mathrm{Tr}\,(\varepsilon R) + \frac{1}{2}\mathrm{Tr}_a\,\mathrm{Tr}_b\,(R_a V_{ab} R_b). \tag{22}$$

Both sides in (22) are given by expansion in powers of I_i. Equating the coefficients in the expansions, we shall obtain numerical parameters H_{ik}, H_{iklm}, ..., as traces over intrinsic states of the nucleus. At this stage we get a macroscopic phenomenological Hamiltonian with its parameters consistently defined from microscopic theory. It is possible to use other consistency conditions, in particular for the total angular momentum, $J_i = \mathrm{Tr}\,(j_i R)$.

7.4. *Applications*

The GDM techniques were successfully applied to several problems.

7.4.1. *Pairing*

This is the simplest example of broken symmetry. The collective band contains the set of the ground states with different particle numbers, ...$N-2, N, N+2$, ..., and N plays the role of conserved momentum. Its intrinsic counterpart is the Pauli matrix τ^3 in the spinor representation (12). The solution of the problem gives the "moment of inertia" for *pair rotation*, that is pair separation energies.

7.4.2. *Translational motion*

In this case the GDM gives a prescription for transformation of the single-particle density matrix ρ (for example, the HF solution) into the GDM R with the conserved linear momentum $\mathbf{P}$. The transformation is given by

$$R = \exp(-i\mathbf{pX})\exp(i(\mu/M)\mathbf{Px})\rho\exp(-i(\mu/M)\mathbf{Px})\exp(i\mathbf{pX}), \tag{23}$$

where $\mathbf{x}$ and $\mathbf{X}$ are the particle and center-of-mass coordinates, respectively, M is the total mass and μ is the reduced mass of the particle with respect to the rest of the nucleus. The self-consistent solution (23) gives the total momentum $\mathbf{P} = \mathrm{Tr}\,(\mathbf{p}R)$ and particle number $A = \mathrm{Tr}\,R = M/m$, where m is the nucleon mass. This exact solution is obtained for the case without pairing and velocity dependent forces. The case with pairing can be solved with accuracy A^{-1}.

7.4.3. *Low spin rotation*

In the case of even-even nuclei considered above as an example, the GDM determines the moments of inertia (all three for triaxial deformation). For axial symmetry, the result for the moment of inertia in the adiabatic approximation coincides with that of the cranking model including the term, caused by the rotational distortion of the MF. Non-adiabatic corrections have improved the agreement with experiments.

7.4.4. *High spin rotation*

In the asymptotic high spin region, $I \gg j$, eq. (11) can be written in the form

$$[h(R') - \mathbf{\Omega}\mathbf{j},\, R'] = i(\mathbf{I} \times \mathbf{\Omega})\frac{\partial R'}{\partial \mathbf{I}}, \tag{24}$$

where $\mathbf{\Omega} = \partial H(\mathbf{I})/\partial \mathbf{I}$, and $\mathbf{I}$ should be treated as a classical quantity. This is, in fact, *time-dependent HF cranking*, since the right hand side, due to the Euler equation $\dot{\mathbf{I}}(t) = \mathbf{I}(t) \times \mathbf{\Omega}(t)$, can be written as $i\dot{R}(t)$.

7.4.5. *Interplay of rotation with intrinsic excitations*

In this case the collective Hamiltonian H in (11) depends not only on angular momentum but also on additional quantum numbers, which characterize other degrees of freedom, for example, vibrations or valence particle(s). The "collective band" in this case is a set of rotational bands based on different band heads, whose characteristics can be chosen as additional quantum numbers of the collective band. The interacting bands built on the RPA β- and γ-phonons were successfully considered.[23]

8. Conclusion and Outlook

The most interesting and promising GDM applications are still waiting for more extensive exploration (collective Hamiltonian for very soft nuclei or other mesoscopic systems with ill-defined shape; enhancement of weak interactions; the limit of many chaotically interacting collective states in relation to random matrix theory etc.) I hope the new results will appear in the near future.

Spontaneous symmetry breaking is a widespread phenomenon in different fields of physics, from condensed matter to elementary particles. In the latter field, for example, a fundamental chiral symmetry breaking is supposed to be restored at the phase transition from hadron matter to quark-gluon plasma. Methods of many-body theory, being developed for one field of physics, are often successfully used in seemingly remote fields; the application of the methods of QED for condensed matter physics was mentioned above.

The present conference gives a marvelous chance for exchange of results, findings and critics as well. That is why I have included in my talk the Generalized Density Matrix approach. It was proved to be successful in nuclear physics. Let me hope that it finds its place in other fields.

To conclude, I would like to express my gratitude to the Feenberg Medal committee for the award and exciting opportunity to take part in this meeting.

References

1. A.B. Migdal and V.M. Galitskii, Sov. Phys. JETP **7**, 96 (1958); V.M. Galitskii, Sov. Phys. JETP **7**, 104 (1958).
2. N.M. Hugenholtz and D. Pines, Phys. Rev. **116**, 489 (1959).
3. S.T. Belyaev. Sov. Phys. JETP **34**, 417 (1958).
4. S.T. Belyaev. Sov. Phys. JETP **34**, 433 (1958).
5. A. Bohr, B.R. Mottelson, and D. Pines, Phys. Rev. **110**, 936 (1958).
6. S.T. Belyaev, Mat. Fys. Medd. Dan. Vid. Selsk. **31**, No. 11 (1959).
7. V. Zelevinsky and A. Volya, Phys. Part. Nucl. **66**, 1829 (2003).
8. P.J. Twin *et al.*, Phys. Rev. Lett. **57**, 811 (1986).
9. T. Byrski *et al.*. Phys. Rev. Lett. **64**, 1650 (1990).
10. S. Flibotte *et al.*, Phys. Rev. Lett. **71**, 4299 (1993).
11. A. Johnson, H. Ryde, and J. Sztarkier, Phys. Lett. **34B**, 605 (1971).
12. R. Waldsworth *et al.*, Phys. Scr. **T88**, 49 (2000).
13. G.B. Hagemann, Eur. Phys. J. A **20**, 183 (2003).
14. A. Bohr and B.R. Mottleson, *Nuclear Structure*, vol. 2 (Benjamin, New York, 1989).
15. D.R. Inglis, Phys. Rev. **103**, 1786 (1956).
16. S. Frauendorf, Nucl. Phys. **A557**, 259c (1993).
17. A. Kerman and A. Klein, Phys. Lett. **1**, 185 (1962); Phys. Rev. **132**, 1326 (1963).
18. A. Klein, Phys. Rev. **138**, B1323 (1965); A. Klein, L. Celenza, and A.K. Kerman, Phys. Rev. **140**, B245 (1965).
19. S.T. Belyaev and V.G. Zelevinsky, Sov. J. Nucl. Phys. **16**, 657 (1973).
20. S.T. Belyaev and V.G. Zelevinsky. Sov. J. Nucl. Phys. **17**, 269 (1973).
21. V.G. Zelevinsky, Nucl. Phys. **A337**, 40 (1980); **344**, 109 (1980).
22. V.G. Zelevinsky, Progr. Theor. Phys. Suppl. **74-75**, 251 (1983).
23. M.I. Shtokman, Sov. J. Nucl. Phys. **22**, 247 (1975).

Keynote Speaker

THE FUTURE LIES AHEAD

P. W. ANDERSON
Department of Physics,
Princeton University,
Princeton, New Jersey 08544

Received 1 December 2004

It was too much of a temptation, when asked to give a talk on the future of our subject, not to use the title of an old LP by the comedian Mort Sahl. The title is at least certainly true, if nothing else I say is.

Many of you may have heard Bob Lauglin's talk at Snowmass a few weeks ago. In it he proposed that we many-body theorists may be the victims of "dark laws" which make it literally impossible to solve some of our problems–specifically he was speaking about high Tc superconductivity. He supposed that there were many possible alternative orderings, so close in energy that we can't possibly resolve their differences. He seems to have been trying to find, in our future, something as mysterious and glamorous as Dark Matter or Dark Energy.

I was reminded of quite a few occasions on which I have heard eminent scientists tolling the death knell over physics–almost always near the end of their productive lives. I won't even use the myth of Lord Kelvin and the fifth decimal place–there was John Bardeen, even, proclaiming in an interview that all the important problems in SSP, as it then was, had been solved–Jim Phillips said wickedly that he meant that all the problems known when JB was a graduate student had been solved. There was Brian Pippard's notorious "Cat and the Cream" proclaiming the death of academic physics in 1960; and Felix Bloch in 1964, as president of the APS, proclaiming that superconductivity hadn't been and wouldn't be solved–a sentiment repeated by John Slater some ten years later. So Bob is not alone in declaiming pessimistic hogwash, though as in the other cases what it no doubt means is that HE won't make much more progress.

I suppose that what this syndrome really means is that each of us has only a finite store of the somewhat irrational optimism that leads us to believe that Nature IS rational and that there IS a solution to every problem if we can only find it.

But enough of what isn't true of the future; What is? Like Newsweek magazine, perhaps I should give you red arrows indicating the state of play in each area. I would suggest that it goes this way:

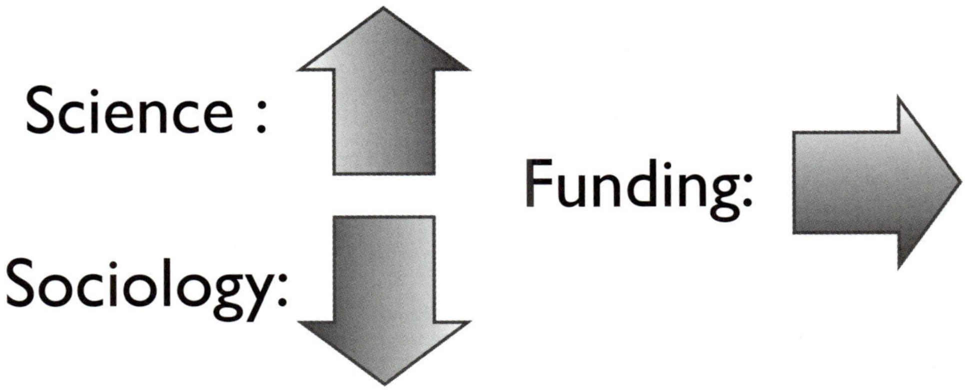

1. Science

Let me start with the good news: the healthy prognosis for our science. To me, the most cheering news is the appearance of many advances in the technology of our subject, advances both in experiment and in theory and calculation methodology. Accurate ARPES, for instance, made its debut just in time to play a big role in high Tc, and the enormous advance represented by the Scienta detector brought a new level of accuracy. Just coming in are two further improvements: laser-excited photoemission with much better energy resolution and greater escape depth; and the automated Dresden facility(though the results appear not yet to be interpreted with adequate sophistication). The ARPES technique has hardly begun to be explored outside of that specialized field, but promising new results have appeared for the cobaltates, for instance.

The development of STM spectroscopy has had a surprising impact on the cuprate field, in spite of the restricted range of materials it has been possible to study; and more is surely on the way.

An exciting new thing is Sawatsky's resonant x-ray scattering, which may open up to us observations of subtle lattice motions such as will accompany the checkerboard pattern seen in STM.

In general, the explosion in nanoscale technology has hardly begun to be exploited–a good example is Ong's recent measurements of fluctuation diamagnetism in cuprates. Even without having detailed knowledge of the field I am always being staggered by what the biophysicists can do with nanotechnology.

Of course I should not omit the wonderful things that are being done with Fermion systems of atoms on optical lattices, and with the availability of a knob to control the size and sign of the interaction.

In theory, the big news is the DMFT (dynamic mean field theory) which gives us a systematic way to deal with the major effects of strong correlations. After nearly 50 years, we are finally able to really understand the Mott transition, for instance, at least in three dimensions, and to model the Kondo volume collapse in cerium. I for one also believe that the sign problem is getting to be less and less of an obstacle to realistic simulations in Fermion problems, and that one of these days we will begin to trust the results, at least in relatively simple systems like the cuprates.

2. Sociology

Here I am afraid that I have some fairly harsh things to say on all sides. Anyone who knows even a little about the history of scientific discovery knows that almost any major discovery is likely to encounter resistance even from the discoverers themselves, who are slow to realize what they have; and almost certainly from referees, rivals–usually the two are the same–and an inevitable coterie of holdouts, of whom those with alternative failed theories are the most stubborn. But in the past the resistance has usually died out more or less inevitably, if necessary due to death or retirement of the most tenacious opponents.

I would suggest that in modern conditions–which John Ziman has aptly named "post-Academic Science", where "academic science" is his description of the way we operated a few decades ago–the inevitability of the victory of obvious scientific truth is a lot less evident. So far, I know of only one subject that has reached the stage where this may happen–namely cuprate superconductivity–but others have come very close to it in the past–a good example being polyacetylene and other polymer conductors.

What causes the problem can be described well by another retro reference–the comic strip Pogo–which produced the immortal phrase, "We has met the enemy and they is us!" It is not in anyone's self-interest to solve the problem!

The experimentalist can ensure his funding, he hopes, and produce a steady stream of papers, by continuing to search for the key effect that is, all by itself, going to make everything clear. (A particularly rewarding game is to search for "THE BOSON which carries the interaction", a chimera based on a naive textbook picture which almost certainly does not apply here). It helps to emphasize, in each of these papers, the important respects in which his or her data differ from those of that other institution which is making the same measurement–even though, under careful examination, the data appear to be equivalent, often identical. (Otherwise one or the other might appear redundant). In his interpretation, he often can enlist the services of his local theorist, or better yet, do his own theory, no matter how amateurish–referees seem to prefer "interpreted" data.

The theorist has even more leeway to further his career at the expense of clarity and simplicity. He can persist bullheadedly with a hypothesis that is plausible to the naive, but physically impossible (That phonons can cause d-wave superconductivity leaps to mind) and find all the loopholes opened up by the fact that it is very hard to prove a negative, especially if your heart isn't in it. Again, the volume of possible papers staggers the mind and enormously enhances the CV–and if lucky, one can enlist some experimentalist whose suggestive but irrelevant data can start a whole fad of imitators. He can base a theory on almost any experimental uncertainty or hypothetical state of matter, so long as it is sufficiently difficult to disprove (see above, but with the added advantage that the state of matter may be experimentally meaningless or nearly so) Wonder of wonders, if he is lucky he can catch the imagination of the editors of some prestigious journal like Nature, so that only papers which refer respectfully to "the stripe theory" (whatever that may be) can appear in that august journal.

It goes without saying that the longer these controversies can be maintained, the more funding has gone into them, the more tenure-track jobs have been awarded, and therefore the more opportunity for new people to start out new fads based on new red-herrings. Who needs a solution to the "problem" posed by the cuprates?– certainly not most of those engaged in the field! Certainly not the funding agencies, whose raison d'etre is the continued confusion; certainly not the deans, chairmen and provosts whose insatiable hunger for overheads is thereby slaked; and of course, certainly not the PI's of all those contracts.

I'm sorry if the above sounds a little bitter. It constitutes my thoughts on reading an extensive file of referee's reports on a recent paper. We had attempted to find the most elementary common ground on which we thought all sensible people could surely agree, and instead we found complete unanimity of rejection but complete diversity in the reasons given, none of which made any sense to us. We could only conclude that there was a universal determination to avoid any appearance of progress towards a real solution, but no other common agreement.

I would caution those who are engaged in pushing the most recent bandwagons for nanotechnology or quantum computing to think a little about the possible consequences of bandwagon formation and hyping the agencies. Did the hype about high Tc do good or harm? Is it really true that more money=better science? Or can one slide into a situation where an unstably growing manifold of nonsense "solutions" leaves those who are actually trying to solve something wandering in the ArXiv, starved for the oxygen of attention?

Strongly Correlated Systems and Phase Transitions

EXACT RESULTS FOR MANY-BODY PROBLEMS USING FEW-BODY METHODS

JOHN CARDY

Rudolph Peierls Centre for Theoretical Physics, 1 Keble Road,
Oxford OX1 3NP, United Kingdom

Received 1 December 2004

Recently there has been developed a new approach to the study of critical quantum systems in 1+1 dimensions which reduces them to problems in one-dimensional Brownian motion. This goes under the name of stochastic, or Schramm, Loewner Evolution (SLE). I review some of the recent progress in this area, from the point of view of many-body theory. Connections to random matrices also emerge.

Keywords: Critical phenomena; conformal field theory; Brownian motion.

1. A Difficult Problem in Many-Body Theory

Consider relativistic charged bosons in D dimensions, with a repulsive interaction, described by a quantum hamiltonian

$$H = \int \left(|\dot{\phi}|^2 + |\nabla\phi|^2 + m_0^2|\phi|^2 + \lambda|\phi|^4\right) d^D x$$

Suppose we are interested in a multi-point correlation function like

$$\langle\Psi|\phi(x'_1, t)\ldots\phi(x'_N, t)\,\phi^\dagger(x_1, 0)\ldots\phi^\dagger(x_N, 0)|\Psi\rangle$$

For $\lambda/m^{3-D} \ll 1$ we could use perturbation theory, but in strong-coupling, critical regime, e.g. as $m \to 0$ at fixed λ, this fails. I want to motivate this talk by first quoting an exact result which was obtained recently. To simplify it, assume we are in a finite system, $0 < x < \ell$, and that t is large, so the correlator factorizes into into two terms. Then each of these takes the form

$$\langle\Psi|\phi(x_1, 0)\ldots\phi(x_N, 0)|0\rangle_N \propto \prod_{j<k}\left|e^{2\pi i x_j/\ell} - e^{2\pi i x_k/\ell}\right|^\beta \tag{1}$$

Remarkably, this has same form as Dyson's circular distribution for the eigenvalues of random unitary matrices. However, in that case the exponent β takes certain discrete values $(1, 2$ or $4)$ depending on the symmetry class of the problem, while in the present case it can take other values: it depends on the number of components n of ϕ and the nature of $|\Psi\rangle$.

This result is one of many which have emerged from an interdisciplinary approach to critical behavior in two (or 1+1) dimensions, which goes under the name of stochastic, or Schramm, Loewner Evolution (SLE) in the mathematics literature,[1] but which is closely related to ideas of conformal field theory (CFT). Both of these subjects are quite technical, so in this talk I can only present the basic ideas and a few specific results.

2. The Equivalent 2d Statistical Mechanics Problem

In the Feynman path integral formulation, the world lines of the particles described by the ϕ field are random curves. For a complex field, these each have an orientation, and, in the sector we are interested in, there are N such curves which begin at $t = 0$ and go off to infinite t. In addition, there are closed loops corresponding to vacuum fluctuations, each weighted $\propto n$. Moreover, in the strong coupling limit in one space dimension we may assume that the curves do not cross.

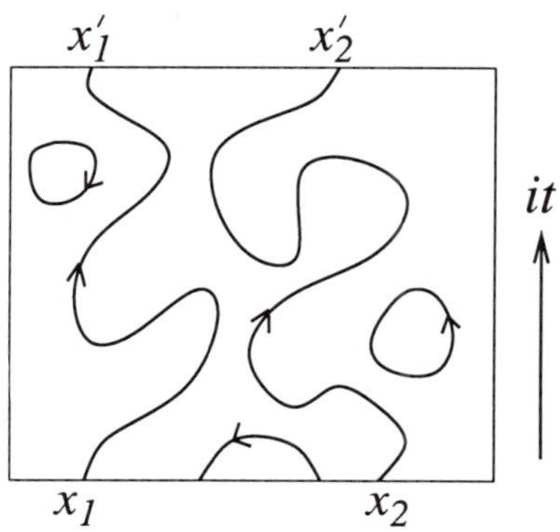

Fig. 1. World lines of charged bosons. In the $\lambda \to \infty$ limit these can be assumed not to cross. Closed loops correspond to vacuum processes. In imaginary time, these curves are also realized as cluster boundaries in 2d classical models.

In 2d classical statistical mechanics, there are many realizations of this gas of non-intersecting loops, for example

- spin cluster boundaries in Ising model ($n = 1$)
- cluster boundaries in percolation (also $n = 1$ but in the so-called dense phase)
- steps on crystal surface near the roughening transition ($n = 2$)
- self-avoiding walks (polymers) ($n \to 0$)

In the scaling limit $m \to 0$ these curves are fractal, and moreover *conformally invariant* in a well-defined sense.

3. Schramm (Stochastic) Loewner Evolution

This is a theory of a *single* curve (particle trajectory) immersed in a sea of vacuum loops. On a lattice, one can in principle think of generating the ensemble of such

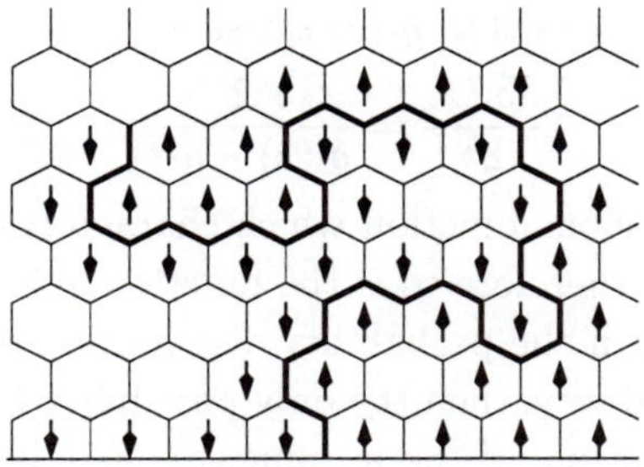

Fig. 2. Exploration process on the lattice.

curves dynamically. For definiteness, consider the upper half plane. Boundary conditions are fixed so that all spins are fixed down on the negative real axis, and up on the positive axis. Instead of generating the whole configuration of all the spins in the upper half plane, and then identifying the cluster boundary which starts at the origin and goes to infinity, we can generate it dynamically by conditioning whether the tip turns left or right depending on the value of the spin in front of it. The relative probabilities of this being up or down depend on the configuration of the curve up to that point.

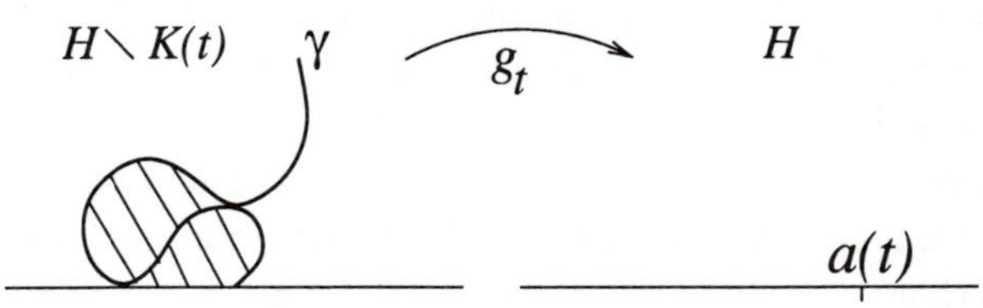

Fig. 3. Schematic drawing of the hull K_t of the curve up to time t, and the conformal mapping which sends the remainder of the upper half plane into the whole of the upper half plane $\mathbf{H}$. The growing tip γ is sent into the point $a(t)$.

SLE is the equivalent process in the continuum limit. Let K_t be the 'hull' (the curve and any regions enclosed by curve) of the curve up to 'time' t. The Riemann mapping theorem asserts that there is a conformal mapping which takes the part of the upper half plane outside K_t, to the whole half plane. This is unique if it is normalized such that

$$g_t(z) \sim z + \frac{2t}{z} + \cdots, \quad \text{as} \quad z \to \infty$$

This also fixes the parametrization of time t. Thus, instead of thinking about the evolution of the tip γ_t of the curve (which may be very irregular), one can think of the evolution of $g_t(z)$, which is quite smooth as long as z is not on the curve. In fact, Loewner's theorem states that $g_t(z)$ satisfies

$$\frac{dg_t(z)}{dt} = \frac{2}{g_t(z) - a(t)},$$

where $a(t)$ is a real continuous function which characterizes the curve. In our case, the curve is random, and therefore so is the function $a(t)$.

Schramm argued that, if the probability distribution of the path is conformally invariant, in a well defined way, then the only possibility for $a(t)$ is one-dimensional Brownian motion. The outline of the argument is as follows (see Fig. 3). Imagine

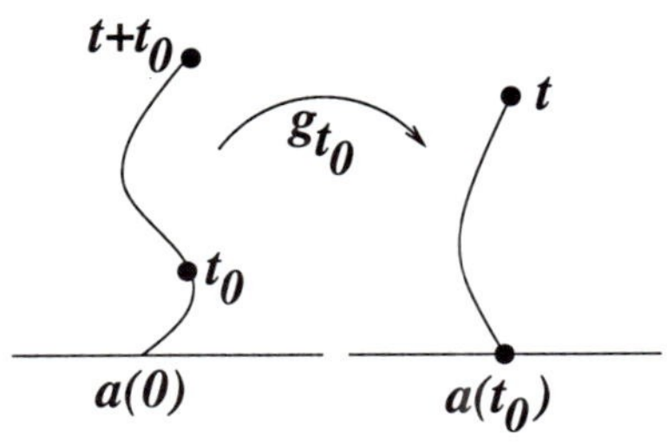

Fig. 4. The image of the curve up, to time $t + t_0$, under g_{t_0}. The law of the the image curve starting at $a(t_0)$ is the same as if we had started from $a(0)$ and evolved for time t.

evolving the path for a time t_0, then for a further time t. Because of conformal invariance, the image under g_{t_0} of the part of the path from t_0 to $t + t_0$ has the same distribution as if we had started the process at $a(t_0)$ and run it for time t. This implies that whatever determines $a(t')$ for $t' > t_0$ is independent of the values of $a(t')$ for $t' < t_0$, i.e. it is a Markov process. Moreover, scale invariance under $z \to \lambda z$, $t \to \lambda^2 t$ implies that $\lambda^{-1}a(\lambda^2 t)$ has the same statistics as $a(t)$. The only solution is Brownian motion: $\dot{a}(t) = \eta(t)$, with $\langle\eta(t')\eta(t'')\rangle = \kappa\delta(t'-t'')$. (A constant drift term is also allowed, but is ruled out if the distribution is to be left-right symmetric.)

This remarkable observation enables many problems in 2d critical systems to be reduced to problems in single-particle 1d Brownian motion. By asking the right question of this stochastic process, in a series of papers Lawler, Schramm and Werner (see 1 for reviews) have been able to compute exactly many of the critical exponents of these systems, as well as other universal properties.

3.1. *Simple properties of SLE*

The first, nontrivial, property is that the inverse mapping $g_t^{-1}(a(t))$ does in fact trace out a mathematical curve. This curve is simple, that is it does not touch itself, for $\kappa \leq 4$, while for $\kappa > 4$ it repeatedly bends back on itself, enclosing whole regions. In fact, the curve eventually swallows the whole upper half plane.

The fractal dimension of the curve is $D_f = 1 + \kappa/8$, for $\kappa \leq 8$, while for $\kappa \geq 8$ $D_f = 1$ and in fact the curve is space-filling: every point in the upper half plane eventually lies on the curve.

The relation between the value of κ and other parameters determining the universality class can be worked out in a number of ways (the simplest involving a comparison of the winding angle distribution of the curves.) For the complex n-component field theories discussed earlier, the relation is

$$n = -2\cos(4\pi/\kappa)$$

The range $2 \leq \kappa \leq 4$ corresponds to the ordinary critical point of the O(n) model, while $4 < \kappa \leq 8$ describes the so-called dense, low-temperature phase which is also critical. So we see that critical Ising model cluster boundaries correspond to $\kappa = 3$, percolation cluster boundaries to $\kappa = 6$, and self-avoiding walks to $\kappa = \frac{8}{3}$, for example.

3.2. *A simple calculation with SLE*

As an example of an SLE result[2] which is not easy to obtain in any other manner, consider a point $\zeta = u + iv$ in the upper half plane. What is the probability $P(u, v; a(0))$ that the curve passes to the right of ζ? Note that, for the Ising model for example, this is not simply related to the average local magnetization, given the boundary conditions, since the question ignores all the islands of opposite magnetization which are surrounded by closed loops. However, by imagining evolving the

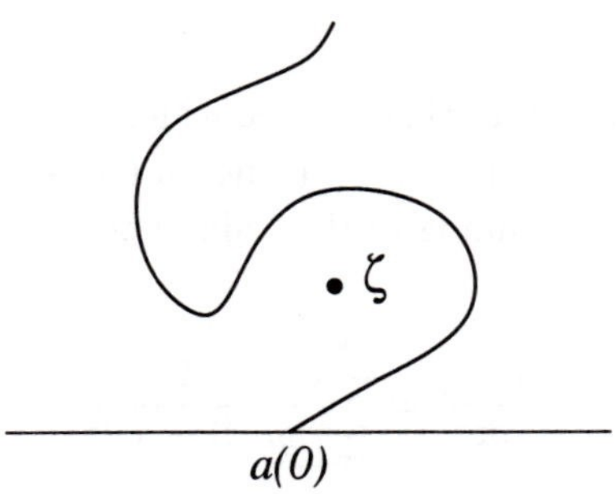

Fig. 5. What is the probability that the curve passes to the right of ζ?

SLE for a short time dt (which does not alter which side of the curve the point falls) we have

$$P(\zeta, \bar{\zeta}, a(0)) = \langle P(g_{dt}(\zeta, \bar{\zeta}), a(dt)) \rangle_{\eta(t');0<t'<dt}$$

where the average on the rhs is over all realizations of the noise up to time dt. Using $\langle a(dt) \rangle = 0$ and $\langle a(dt)^2 \rangle = \kappa dt$, we find

$$\left(\frac{u}{u^2 + v^2} \frac{\partial}{\partial u} - \frac{v}{u^2 + v^2} \frac{\partial}{\partial v} + \frac{\kappa}{2} \left(\frac{\partial}{\partial a(0)} \right)^2 \right) P = 0$$

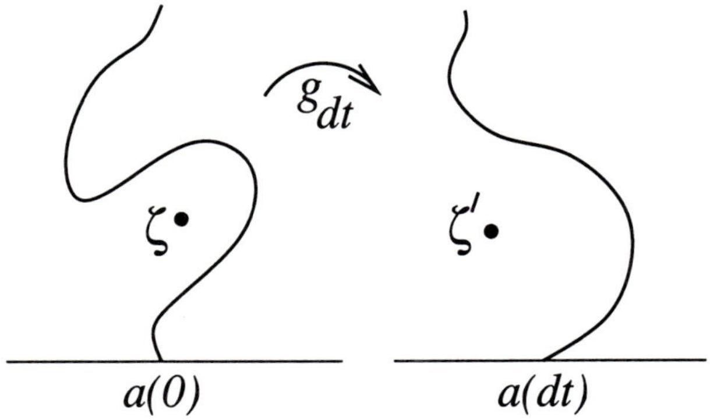

Fig. 6. Image of the previous figure under g_{dt}.

Using the fact that P can in fact depend only on the angle subtended by ζ at $a(0)$, reduces this to an ordinary differential equation whose solution is simple.

An important observation, first made by Bauer and Bernard,[3] is that the differential equations one gets this way are the same as those coming from boundary conformal field theory (CFT). Thus SLE also gives an alternative way of understanding CFT.

4. Multiple Curves

An obvious generalization[4] of all this is to consider multiple curves, or world-lines, beginning at points $a_j(0)$, $j = 1, \ldots, N$ on the real axis. This will bring us back to the problem posed at the beginning of this talk. The generalization of the Loewner equation is

$$\frac{dg_t(z)}{dt} = \sum_{j=1}^{N} \frac{2}{g_t(z) - a_j(t)}$$

but what are the dynamics of the $a_j(t)$? This can be answered by considering what happens if we allow each curve to advance in turn, in the background of the others. In this way we get

$$\dot{a}_j = \sum_{k \neq j} \underbrace{\frac{2}{a_j - a_k}}_{k \text{ advances}} + \underbrace{\eta_j}_{j \text{ advances}} \tag{2}$$

The first term on the rhs comes from applying the Loewner equation to the point a_j during the time when the kth curve advances, while the second term comes from the random motion of a_j when the jth curve advances. There appears to be some arbitrariness in all of this: why, for example, should the time parameter be the same for all the curves? This can be settled by an appeal to CFT.[5] The striking thing about (2) is it is the same as Dyson's Brownian motion[6] for eigenvalues of random

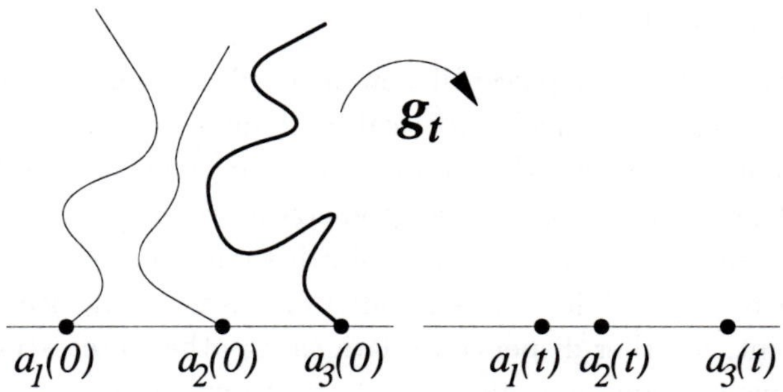

Fig. 7. Multiple curves, and their image under the generalized Loewner map.

hermitian matrices. If the individual elements of an $N \times N$ hermitian matrix undergo uncorrelated Brownian motion, then its eigenvalues satisfy (2), with, however, a specific value of κ. Perhaps this is just a coincidence: (2) is the only such equation which satisfies such requirements as invariance under rescaling $t \to \lambda^2 t$, $a_j \to \lambda a_j$.

4.1. *Circular version*

Since the points repel each other in (2), the process does not reach equilibrium. Instead consider the equivalent problem in which the real line is replaced by a circle, $x = 2\pi\theta/\ell$, the upper half plane by a semi-infinite cylinder.(2) then becomes

$$\dot{\theta}_j = \sum_{k \neq j} \cot\left(\frac{\theta_j - \theta_k}{2}\right) + \eta_j$$

This is Dyson's Brownian motion for the eigenvalues of unitary matrices. Now as $t \to \infty$, the joint probability distribution of the $\{\theta_j\}$ tends to the equilibrium

$$P_{\text{eq}} \propto \prod_{j<k} |e^{i\theta_j} - e^{i\theta_k}|^\beta \qquad (3)$$

with $\beta = 4/\kappa$. After some further arguments, this leads to the result stated in (1). In fact, these are slightly subtle, since it turns out that the correlation function is not quite invariant under the process. This leads to a shift $\beta \to 8/\kappa$. However the exponent in (1) is also changed, to $2/\kappa$.[5]

Note that the case $n = 2$ (a single charged boson) corresponds to $\beta = 1$. The result is then the wave function for free fermions. This is what would be expected for *non-relativistic* hard core bosons, but in our case the bosons are relativistic: their world lines can go backward in time. But the mutual repulsion this produces makes the correlation function look just like that of free fermions.

The generator of Dyson's process is related by a similarity transform to the famous N-body Calogero-Sutherland (CS) hamiltonian. This means that the equilibrium distribution (3) is simply related to its ground state wave function. It turns out that other eigenvalues of this hamiltonian give bulk critical exponents for this problem, for example related to the probability that N lines meet near one point.[5]

5. Summary and Outlook

We have seen that SLE is a powerful new method of analyzing 1D (relativistic) quantum models and 2d classical statistical mechanics models at critical points. It takes a very geometrical, rather than a quantum field, point of view of the problems. It relies on conformal invariance of this geometrical picture.

Although it gives many old and new results, it should not be seen as all-powerful. In particular, it appears difficult to extend these ideas to the scaling region away from critical point, to other dispersion relations, to other interesting and unsolved problems in quenched random systems, and, most importantly, to $D > 1$.

Acknowledgments

This work was supported in part by the EPSRC under Grant GR/R83712/01, and was completed while the author was a member of the Institute for Advanced Study. He thanks the School of Mathematics and the School of Natural Sciences for their hospitality. This part of the work was supported by the Ellentuck Fund.

References

1. This was first proposed in O. Schramm, *Israel J. Math.* **118**, 221 (2000). For mathematical reviews, see W. Werner, *Random planar curves and Schramm-Loewner evolutions*, to appear (Springer Lecture Notes), math.PR/0303354; G. Lawler, *Conformally Invariant Processes in the Plane*, in preparation, http://www.math.cornell.edu/~lawler/book.ps. For a review aimed at theoretical physicists, see W. Kager and B. Nienhuis, *J. Stat. Phys.* **115**, 1149 (2004).
2. O. Schramm, math.PR/0107096.
3. M. Bauer and D. Bernard, *Comm. Math. Phys.* **239**, 493 (2003); *Phys. Lett. B* **543**, 135 (2002); *Phys. Lett. B* **557**, 309 (2003); math-ph/0305061.
4. J. Cardy, *J. Phys. A* **36**, L379 (2003); erratum, *J. Phys. A* **36**, 12343 (2003).
5. J. Cardy, *Phys. Lett. B* **582,** 121 (2004).
6. F. Dyson, *J. Math. Phys.* **3**, 1191 (1962).

QUANTUM MATTERS: PHYSICS BEYOND LANDAU'S PARADIGMS

T. SENTHIL

Department of Physics, Massachusetts Institute of Technology,
Cambridge, Massachusetts 02139
senthil@mit.edu

Received 1 December 2004

Central to our understanding of quantum many particle physics are two ideas due to Landau. The first is the notion of the electron as a well-defined quasiparticle excitation in the many body state. The second is that of the order parameter to distinguish different states of matter. Experiments in a number of correlated materials raise serious suspicions about the general validity of either notion. A growing body of theoretical work has confirmed these suspicions, and explored physics beyond Landau's paradigms. This article provides an overview of some of these theoretical developments.

Keywords: Quantum phase transitions; fractionalization; critical phenomena; Landau-Ginzburg-Wilson approach.

1. Introduction

Much of our understanding of quantum many electron physics rests on two central ideas which may be traced to Landau. The first is the notion that the electron retains its integrity as a quasiparticle excitation above the quantum ground state of the many particle system. This notion underlies Landau's celebrated Fermi Liquid theory of metals but is also shared by many familiar phases of matter (band insulators, BCS superconductors, spin density waves, ...). This is true even if microscopically the electrons interact reasonably strongly with each other. The other important idea of Landau is that of the order parameter to classify and distinguish phases of matter. Closely related is the notion of spontaneously broken symmetry - indeed the Landau order parameter quantifies the amount of symmetry breaking in any ordered phase. The concept of the order parameter plays an important role in phase transition theory. The universal critical singularities at second order phase transitions are usually attributed to the long wavelength fluctuations of the order parameter degrees of freedom. When combined with general renormalization group ideas this gives a sophisticated theoretical framework — often known as the Landau-Ginzburg-Wilson(LGW) paradigm — for describing phase transition phenomena.

Both of these two notions - the integrity of the electron and the Landau order parameter - are so fundamental that they are routinely taught at the early stages of a solid state physics education. Remarkably in the last several years, a number of experimental developments have challenged the general applicability of either of these two basic notions to quantum condensed matter. The best known is the phenomenon of the quantum Hall effect that occurs in a two dimensional electron gas in high magnetic field.[1] The quantum Hall systems are striking and well-established examples of the violation of both of Landau's paradigms. Indeed the electron does not in general survive as a quasiparticle[2] in fractional quantum Hall states nor is the order in such a state captured by a local Landau order parameter.[3]

Since the discovery and explanation of the quantum Hall effect, a number of other remarkable phenomena have been found in other correlated electron systems which are poorly understood. Perhaps the most notorious is the problem of high temperature superconductivity but there are many others as well. Examples include a host of "non-fermi liquid" phenomena in the rare-earth intermetallics known as the heavy fermions,[4] various kinds of materials near Mott metal-insulator transitions, etc. Many of these long standing problems in condensed matter theory seem not to yield to any conventional thinking. They presumably require new methods of attack, new languages, and perhaps even new conceptual advances. At any rate these experimental discoveries certainly lead to the suspicion that Landau's paradigms may breakdown in serious ways in correlated many electron systems. In the last few years, this suspicion has been strikingly confirmed by a variety of theoretical advances that unquestionably demonstrate the inadequacies of Landau's paradigms for a general understanding of correlated matter.

This article provides an overview of these theoretical developments. We begin by briefly discussing phases of quantum matter that do not fit in with Landau's ideas. We then discuss very recent work on the breakdown of the Landau paradigm at zero temperature 'quantum' phase transitions.

2. Breakdown of Landau Paradigms in Correlated Quantum Phases

As mentioned above the fractional quantum Hall effect provides a well-established (and by now old) example of a correlated phase that violates Landau's paradigms. An even older example is provided by one dimensional systems such as (half-integer) quantum spin chains or polyacetylene.[5] A more modern realization of one dimensional physics occurs in the Carbon nanotubes.[6] In these examples, the electron does not retain its integrity as a quasiparticle excitation. Rather the excitations have quantum numbers that are fractions of those of the electron.

One of the most interesting theoretical developments in the last several years is the realization that such 'broken electron' phenomena are *not* restricted to such extreme situations as one dimension or two dimensions in strong magnetic fields. Indeed it has become clear that electrons can break apart in regular solids with strong electron-electron interaction in any spatial dimension. The electron does not survive as a quasiparticle in such phases of matter — instead there are excitations with quantum numbers that are fractions of those of the electron.

A great deal has been learnt about these 'fractionalized' phases in $d \geq 2$ at zero magnetic field theoretically. The structure of the excitation spectrum (or more generally the structure of the low energy effective field theory) has been elucidated.[7,8,9,10,11,12] These phases have a certain kind of 'order' that is not captured by a local Landau order parameter. Rather the ordering is a global property of the many electron ground state wavefunction[13,8,14] — often referred to as 'topological order'. This kind of order generalizes and is indeed distinct from the old notion of spontaneously broken symmetry. Several concrete and simple microscopic models which display these phenomena exist in both two[15,16,17,18,19,20,21,22] and three spatial dimensions.[20,23,24,25] Prototypical ground state wavefunctions for a number of such fractionalized states can be written and explicitly shown to possess properties (such as topological order) expected on general grounds.[13,26] Finally there even exist some ideas on how to directly detect certain kinds of topological order in experiments.[14,27,28]

All of this is spectacular theoretical progress — much of it happened in the last five or so years and builds on important ideas and results[29,13,7,8] from the early days of high-T_c theory. In particular many of the theoretical criticisms levelled against these ideas have now been satisfactorily answered. However there still is no unambiguous identification of such broken electron phenomena in experiments other than the previously established instances (FQHE and $d = 1$).

Where else might it happen? The theoretical understanding provides some hints. It has long been appreciated that frustrated quantum magnets may be a good place to look for such physics. It has also become clear that other promising candidates are not-so-strongly correlated materials. This may be seen explicitly in some of the microscopic models showing fractionalization where it appears in intermediate correlation regimes where neither kinetic nor potential energy overwhelmingly dominates the other.[20] Further support is provided by the observation that in spin systems fractionalization is promoted by multi-particle ring exchange terms[30,10,18]ch terms become increasingly important for the spin physics of Mott insulators as one moves away from the very strong interaction limit (decreasing U in a Hubbard model description). Thus Mott insulators that are not too deeply into the insulating phase or quantum solids such as He-3 or He-4 near melting may be good places to look as well.

3. Breakdown of Landau Paradigms at Quantum Phase Transitions

We now turn to the breakdown of Landau's paradigms at zero temperature 'quantum' phase transitions. That this might happen was originally hinted at by various distinct kinds of observations in the literature. First as reviewed in the previous section, Landau order parameters do not necessarily capture the true order in quantum phases. Then it is quite natural that transitions out of such phases are not described by Landau ideas either. For instance continuous transitions exist between distinct quantum Hall states which clearly cannot be described in terms of simple order parameter fluctuations a'la Landau. But what about transitions out of of phases in which Landau order parameters do capture the order? Here at least one might have hoped for Landau ideas on phase transitions to work. We now review recent work showing that even in this case the Landau paradigm breaks down.

The possibility of such a breakdown is suggested by two different observations. The first is in numerical calculations on various quantum transitions that see a direct second order quantum phase transition between two phases with different broken symmetry characterized by two apparently independent order parameters.[31,32] This is in general forbidden within the Landau approach to phase transitions except at special multicritical points. A similar phenomenon is also seen in experiments[33] on the heavy fermion compound UPt_3. At low temperatures, this is a superconductor (believed to be triplet paired). Upon doping Pd into the Pt site, the superconductivity very quickly disappears and is replaced instead by an antiferromagnetic metal. Within the resolution of the existing experiments these two different kinds of order (superconductivity and antiferromagnetism) seem to be separated by a direct second order transition — within Landau order parameter theory this too would be a special accident. However the surprising frequency with which such 'Landau-forbidden' quantum transitions show up suggests a reexamination of the validity of the Landau paradigm itself.

A second and perhaps more important reason to suspect the general validity of the Landau paradigm comes from a number of fascinating experiments probing the onset of magnetic long range order in the heavy fermion metals.[4] Remarkably the behavior right at the quantum transition between the magnetic and non-magnetic metallic phases is very strikingly different from that of a fermi liquid. The natural assumption is to attribute the non-fermi liquid physics to the universal critical singularities of the quantum critical point. Within the Landau paradigm these will be due to long wavelength long time fluctuations of the natural magnetic order parameter. In other words the hope is that Landau's ideas on phase transitions may perhaps be used to kill Landau's theory of Fermi Liquids. However theories associating the critical singularities with fluctuations of the natural magnetic order parameter in a metallic environment[34] seem to have a hard time explaining the observed non-fermi liquid phenomena. This failure once again fuels the suspicion that perhaps the Landau approach to phase transitions is incorrect. Specifically other phenomena such as the possible loss of Kondo screening of local moments may contribute[35] to and perhaps even dominate the critical singularities.[36] This kind of thinking — particularly the latter possibility — is clearly outside the LGW framework for critical phenomena. In other words *the Landau order parameter (even if present) may distract from the fluctuations responsible for the true critical behavior.*

These suspicions have been strikingly confirmed in recent theoretical work[37,38] on quantum phase transitions in insulating magnets in two spatial dimension. As usual insulating magnets provide a good theoretical laboratory to study phase transition phenomena. A number of results have been found which quite clearly demonstrate the failure of LGW theory at certain (but not all) quantum phase transitions.

In all the examples studied so far the critical phenomenology is instead apparently most conveniently described in terms of objects that carry fractional quantum numbers and which interact with each other through emergent gauge forces. These fractional objects do not necessarily exist (as good excitations) in the two phases but become useful degrees of freedom at the quantum critical point. This kind of phenomenon has been dubbed 'deconfined' quantum criticality — with a sharp and specific meaning of the term 'deconfined'.

Consider a spin-1/2 antiferromagnet on a two dimensional square lattice described by a Hamiltonian of the general form

$$H = J \sum_{<rr'>} \vec{S}_r.\vec{S}_{r'} + \cdots. \tag{1}$$

The $\vec{S}_r$ are spin-1/2 operators and $J > 0$ is the nearest neighbour exchange constant. The ellipses represent other interactions such as a diagonal exchange or multiparticle ring exchange that may be tuned to drive phase transitions. In the absence of these extra terms the ground state is known to have long ranged Neel order. The corresponding order parameter is a vector in spin space $\vec{N} \sim (-1)^{(x+y)}\vec{S}_r$. For suitable choices of the extra terms it is expected that the ground state will not have long range Neel order even at zero temperature. The simplest of such 'quantum paramagnets' are states known as valence bond solids(VBS) — see Fig. 1 In a cartoon of such states each spin forms a singlet valence bond with one of its neighbours. The resulting dimers stack up in some particular pattern in the VBS ground state. The resulting state clearly has spin rotation symmetry but the pattern of dimer ordering breaks various lattice symmetries. Clearly the order parameter for the VBS state is a spin singlet that transforms non-trivially under the lattice space group operations — it is readily constructed out of the bond energy operators $\vec{S}_r.\vec{S}_{r'}$. The elementary spin-carrying excitations in this phase are gapped spin triplet particles.

In a naive Landau description of the two phases that focuses only on the low energy order parameters, a direct second order transition is not expected except at fine tuned multicritical points. However this naive expectation has been argued to be incorrect. A generic second order transition is possible between these two phases with different broken symmetries. The resulting critical theory is however unusual and *not* naturally described in terms of the order parameter fields of either phase.

The key reason behind this violation of naive Landauesque expectation lies in the observation[39,40] that topological defects in either order parameter carry non-trivial quantum numbers. In particular the defects in one order parameter transform in the same way under the microscopic symmetries as the order parameter for the other phase. Thus when the defects in, say the Neel vector configuration, proliferate they destroy long range Neel order. However the non-trivial quantum numbers they carry induces VBS order in the resulting paramagnet.[41]

In the theory of Ref. 37, 38 the natural description of the transition is in terms of spin-1/2 "spinon" fields z_α ($\alpha = 1, 2$ is a spinor index). The Néel order parameter

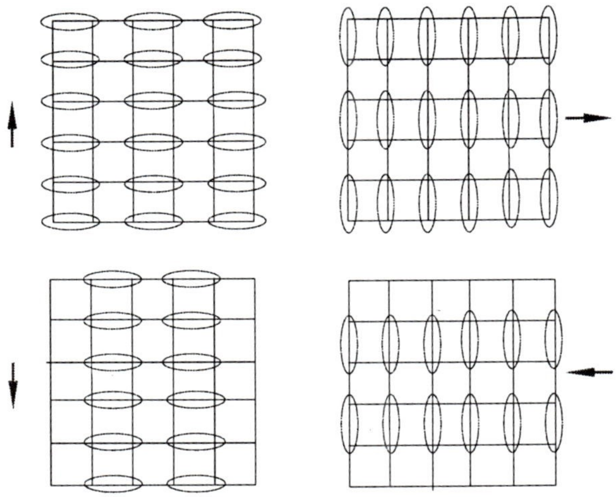

Fig. 1. Schematic picture of a columnar VBS state showing the four degenerate ground states. The encircled lines represent the bonds across which the spins are paired into a valence bond. The four ground states are associated with four different orientations of a Z_4 clock order parameter.

is bilinear in the spinons:

$$\vec{N} \sim z^\dagger \vec{\sigma} z. \tag{2}$$

Here $\vec{\sigma}$ is the usual vector of Pauli matrices and multiplication of the spinor index is implied. The fields z_α create single spin-1/2 quanta, "half" that of the spin-1 quanta created by the Néel field $\vec{N}$. The analysis of Ref. [37,38] shows that the correct critical field theory has the action $\mathcal{S}_z = \int d^2r d\tau \mathcal{L}_z$, and

$$\mathcal{L}_z = \sum_{a=1}^{2} |\left(\partial_\mu - i a_\mu\right) z_a|^2 + s|z|^2 + u\left(|z|^2\right)^2 + \kappa \left(\epsilon_{\mu\nu\kappa}\partial_\nu a_\kappa\right)^2, \tag{3}$$

This is distinct from both the $O(3)$ universality class in $D = 3$ (as might have been expected based on the Neel order parameter) or the Z_4 universality class (as expected from the Z_4 VBS order parameter). The distinction with the $O(3)$ universality class may be a bit puzzling to field theorists familiar with the CP^1 description of the $O(3)$ non-linear sigma model — the crucial point is that the gauge field in Eqn. 3 above is non-compact. As explained in Ref. 42 with a non-compact gauge field this model does not describe the usual $O(3)$ ordering transition in $D = 3$ — rather it describes the transition[43] in $O(3)$ models wehere 'hedgehog' defects have been suppressed by hand.[44] Ref. 42 also contains detailed numerical calculations of critical exponents for the transition in the model Eqn. 3. The non-compactness of the gauge field leads to an extra emergent conservation law (conserved gauge flux) that helps give precise meaning to the notion of deconfinement at the critical point. This conservation law emerges only at the critical point and does not obtain away from it in either phase.

The critical behavior at this transition is strikingly anamolous — indeed it may be viewed as the moral equivalent of 'non-fermi liquid' behavior in this insulating context. For instance the magnon spectral function is extremely broad when compared to other quantum transitions - the exponent η is estimated[42] to be ≈ 0.6, bigger by an order of magnitude as compared to the conventional $O(3)$ fixed point. This may roughly be understood as being due to the decay of magnons into the spinon degrees of freedom. A number of other interesting properties — such as the presence of more than one diverging length/time scale — have also been found.[37,38]

The Neel-VBS transition is not the only example of this kind of quantum phase transition. A number of other transitions in quantum antiferromagnetism have been shown to have many similarities with the phenomena described above. These include the transition from the VBS state to a gapped 'spin liquid' paramagnet[38] and quantum transitions between two different patterns of VBS ordering on certain lattices[45,46] (for instance in a bilayer honeycomb lattice). Further deconfined critical *phases* described by gapless Dirac-like fermionic spin-1/2 objects coupled to an emergent non-compact $U(1)$ gauge field have been shown to exist as stable quantum phases[47] in two space dimensions. Thus it appears that the phenomenon of deconfined quantum criticality is reasonably common in two dimensional quantum magnets.

Easy plane versions of quantum spin-1/2 models have also been examined and shown to have Landau-forbidden transitions and asociated deconfined quantum critical points.[37,38] These may also be fruitfuly viewed as superfluid-insulator transitions of bosons at half-filling on the square lattice. The case of bosons at a general commensurate filling p/q has been examined recently[48] — again the topological defects have been shown to carry non-trivial quantum numbers which in turn leads to non-trivial order in the insulating phase. Direct second order Landau-forbidden transitions seem possible for a number of special fillings.

4. Conclusions

The developments discussed above provide a theoretically important zeroth order answer to the basic question posed by experiments in modern correlated electron physics: Can Landau's ideas breakdown in quantum matter more generally than in one dimension or the quantum Hall effect? While the theoretical progress at this basic level has been dramatic, we do not at present know what role, if any, it will play in understanding existing experiments on materials such as the cuprates or the heavy fermions. Nevertheless the intuition gleaned from these results will hopefully suggest ways of thinking correctly about such experimental problems.

Acknowledgments

It is a pleasure to thank my many collaborators on these matters - a partial list is P. Ghaemi, M. Levin, M. Hermele, O. Motrunich, A. Vishwanath, C. Lannert, D. Ivanov, L. Balents, S. Sachdev, Matthew Fisher, P.A. Lee, N. Nagaosa, and X.-G.

Wen. This research is supported by the National Science Foundation grant DMR-0308945, the NEC Corporation, the Alfred P. Sloan Foundation, and The Research Corporation.

References

1. See, for example, Perspectives in Quantum Hall Effects, edited by S. DasSarma and A. Pinczuk (Wiley, New York, 1997).
2. R. B. Laughlin, Phys. Rev. Lett. 50, 1395 (1983); V.J. Goldman and B. Su, Science 267, 1010 (1995); R. de-Picciotto et al., Nature (London) 389, 162 (1997) L. Saminadayar, D. C. Glattli, Y. Jin, and B. Etienne Phys. Rev. Lett. 79, 2526-2529 (1997)
3. X. G. Wen and Q. Niu, Phys. Rev. B 41, 9377 (1990) and references therein.
4. H. v. Lhneysen et al., Phys. Rev. Lett. 72, 3262 (1994); Mathur, N. D. et al. Magnetically mediated superconductivity in heavy fermion compounds. Nature 394, 39-43 (1998); J. Custers et al., Nature (London) 424, 524 (2003)
5. For reviews see A. Auerbach, Interacting Electrons and Quantum Magnetism (Springer-Verlag, New York, 1994); A. J. Heeger, S. Kivelson, J. R. Schrieffer, and W. -P. Su Rev. Mod. Phys. 60, 781-850 (1988).
6. See for instance M. Bockrath et al., Nature (London) 397, 598 (1999)
7. N. Read and S. Sachdev, Phys. Rev. Lett. 66, 1773 (1991)
8. X.G. Wen, Phys. Rev. B 44, 2664 (1991)
9. L. Balents, M. P. A. Fisher, and C. Nayak Phys. Rev. B 60, 1654 (1999)
10. T. Senthil and M. P. A. Fisher, Phys. Rev. B 62, 7850-7881 (2000)
11. M. Freedman, C. Nayak, K. Shtengel, K. Walker, and Z. Wang (2003), cond-mat/0307511
12. M.A. Levin and X.-G. Wen, cond-mat/0404617
13. N. Read and B. Chakraborty, Phys. Rev. B 40, 7133 (1989)
14. T. Senthil and M. P. A. Fisher, Phys. Rev. Lett. 86, 292 (2001); Phys. Rev. B 63, 134521 (2001)
15. A. Y. Kitaev, quant-ph/9707021; Ann. Phys. (N.Y.) 303, 2 (2003)
16. R. Moessner and S. L. Sondhi, Phys. Rev. Lett. 86, 1881 (2001)
17. G. Misguich et al., Phys. Rev. Lett. 89, 137202 (2002)
18. L. Balents, M. P. A. Fisher, and S. M. Girvin Phys. Rev. B 65, 224412 (2002)
19. T. Senthil and O. Motrunich Phys. Rev. B 66, 205104 (2002)
20. O. I. Motrunich and T. Senthil Phys. Rev. Lett. 89, 277004 (2002)
21. O. I. Motrunich, Phys. Rev. B 67, 115108 (2003)
22. Xiao-Gang Wen, Phys. Rev. Lett. 90, 016803 (2003)
23. Xiao-Gang Wen, Phys. Rev. B 68, 115413 (2003); Michael Levin et al., Phys. Rev. B 67, 245316 (2003)
24. M. Hermele, M.P.A. Fisher, L. Balents, Phys. Rev. B 69, 064404 (2004)
25. R. Moessner and S. Sondhi, Phys. Rev. B 68, 184512 (2003)
26. D. A. Ivanov and T. Senthil Phys. Rev. B 66, 115111 (2002); A. Paremekanti et al, cond-mat/0405353
27. T. Senthil and M.P.A. Fisher, Phys. Rev. B 64, 214511 (2001)
28. L. B. Ioffe et al, Nature (London) 415, 503 (2002)
29. P. W. Anderson, Science 235, 1196 (1987); S. A. Kivelson, D. S. Rokhsar, and J. P. Sethna, Phys. Rev. B 35, 8865 (1987).
30. G. Misguich et al., Phys. Rev. B 60, 1064 (1999)
31. F. F. Assaad, M. Imada, and D. J. Scalapino, Phys. Rev. Lett. 77, 4592 (1996)
32. A. W. Sandvik et al. , Phys. Rev. Lett. 89, 247201 (2002)

33. A. de Visser et al, Phys. Rev. Lett. 85, 3005-3008 (2000)
34. M. T. Beal-Monod and K. Maki, Phys. Rev. Lett. 34, 1461 (1975); J. A. Hertz, Phys. Rev. B 14, 1165 (1976); T. Moriya, Spin Fluctuations in Itinerant Electron Magnetism, Springer-Verlag, Berlin (1985); G. G. Lonzarich and L. Taillefer, J. Phys. C 18, 4339 (1985); A. J. Millis, Phys. Rev. B, 48, 7183 (1993).
35. P. Coleman et al, J. Phys:Condens. Matt. 13, 723 (20010; Q. Si et al, Nature 413, 804 (2001) and Phys. Rev. B 68, 115103 (2003).
36. T. Senthil, Matthias Vojta, and Subir Sachdev Phys. Rev. B 69, 035111 (2004), and cond-mat/0409033
37. T. Senthil, A. Vishwanath, L. Balents, S. Sachdev, and M. P. A. Fisher, Science 303, 1490 (2004).
38. T. Senthil, L. Balents, S. Sachdev, A. Vishwanath, and M. P. A. Fisher, Phys. Rev. B 70, 144407 (2004)
39. F. D. M. Haldane, Phys. Rev. Lett. 61, 1029 (1988)
40. M.A. Levin and T. Senthil, cond-mat/0405702
41. N. Read and S. Sachdev, Phys. Rev. B 42, 4568 (1990)
42. O. I. Motrunich and A. Vishwanath, Phys. Rev. B 70, 075104 (2004)
43. M. Kamal and G. Murthy, Phys. Rev. Lett. 71, 1911 (1993)
44. M.-h. Lau and C. Dasgupta, Phys. Rev. B 39, 7212 (1989)
45. A. Vishwanath, L. Balents, and T. Senthil, Phys. Rev. B 69, 224416 (2004).
46. E. Fradkin, D. A. Huse, R. Moessner, V. Oganesyan, and S. L. Sondhi, Phys. Rev. B 69, 224415 (2004); E. Ardonne, P. Fendley, and E. Fradkin, Ann. Phys. (N.Y.) 310, 493 (2004).
47. M. Hermele, T. Senthil, Matthew P. A. Fisher, Patrick A. Lee, Naoto Nagaosa, Xiao-Gang Wen, cond-mat//0404751
48. Leon Balents, Lorenz Bartosch, Anton Burkov, Subir Sachdev, K. Sengupta, cond-mat/0408329

MICROSCOPIC CALCULATIONS OF QUANTUM PHASE TRANSITIONS IN FRUSTRATED MAGNETIC LATTICES

RAYMOND F. BISHOP

Department of Physics
University of Manchester Institute of Science and Technology (UMIST)
P.O. Box 88, Manchester M60 1QD, United Kingdom
raymond.bishop@manchester.ac.uk

SVEN E. KRÜGER

Department of Electrical Engineering
IESK, Cognitive Systems, Universität Magdeburg
PF 4120, 39016 Magdeburg, Germany

Received 1 December 2004

It is nowadays acknowledged that the coupled cluster method (CCM) provides one of the most powerful and most widely used of all *ab initio* techniques of microscopic quantum many-body theory. It has been applied to a broad range of both finite and extended physical systems defined on a spatial continuum, where it has generally yielded numerical results which are among the most accurate available. This widespread success has spurred many recent applications to quantum systems defined on a lattice. We discuss here a typical example of a two-dimensional spin-half Heisenberg magnet with two kinds of competing nearest-neighbour bonds. We show how the CCM can successfully describe the influence of strong quantum fluctuations on the zero-temperature phases and their quantum phase transitions. The model shows how the CCM can successfully describe the effects of competition between magnetic bonds with and without the presence of frustration. The frustrated case is particulary important since many other methods, including quantum Monte Carlo simulations, typically fail in this regime.

Keywords: Coupled cluster method; quantum phase transitions; magnetic lattices; frustration; bond competition.

1. Introduction

The *coupled cluster method* (CCM)[1] has become one of the most pervasive and most powerful of all *ab initio* formulations of microscopic quantum many-body theory. It has been widely applied to many diverse physical and chemical systems defined on a spatial continuum,[2] where it has yielded numerical results which are usually among the best available for these systems. This widespread success has spurred recent further applications to similar strongly-correlated quantum-mechanical systems defined on an extended regular spatial lattice.

Such spin- and electron-lattice systems have become the subject of intense theoretical study. They include many examples of systems characterized by novel ground states which display *quantum order* in some region of the Hamiltonian parameter space, delimited by critical values which mark the corresponding *quantum phase transitions.* The critical phenomena in these quantum-mechanical models often differ profoundly from those of their classical counterparts, and the subtle correlations present usually cannot sensibly be treated by such standard many-body techniques as perturbation theory or mean-field approximations. One of the key challenges for modern quantum many-body theory is to develop and exploit microscopic techniques capable of handling both these novel and more traditional systems. Our recent work[3] shows that one such method which is capable of bridging this divide is the CCM.

In particular, we have shown how the systematic inclusion of multispin correlations for a wide variety of quantum spin-lattice problems can be very efficiently implemented with the CCM.[4] The method is not restricted to bipartite lattices or to non-frustrated systems, and can thus deal with problems[3–13] where the quantum Monte Carlo (QMC) techniques would be faced with the infamous "minus-sign problem". We note that the method is also easily extended to models of strongly interacting electrons on lattices, such as the Hubbard model.[14]

In Sec. 2 we first very briefly review the CCM itself as a universal tool of quantum many-body theory. We then discuss an illustrative example from among many applications which have already been made to quantum spin-lattice systems. The model itself, which contains two types of competing nearest-neighbour interactions, is first described in Sec. 3, where its classical ground state is also discussed. The model is particularly interesting because, *inter alia*, it exhibits competition between magnetic order and dimerization. As in all other cases the CCM may readily be implemented for this model to high orders using computer-algebraic techniques. Results are presented in Sec. 4, where critical comparisons are made with the results from such other competing techniques as exact diagonalisations of small lattices, and (both linear and second-order renormalized) spin-wave theory. We conclude with a discussion of the results in Sec. 5.

2. The Coupled Cluster Method

The starting point for any CCM calculation (and see the overviews in Refs. [1, 2] for more details) is the choice of a normalized model or reference state $|\Phi\rangle$, together with a set of mutually commuting multispin creation operators C_I^+ which are defined over a complete set of many-body configurations I. The operators C_I are the multispin destruction operators and are defined to be the Hermitian adjoints of the C_I^+ operators. We choose $\{|\Phi\rangle; C_I^+\}$ in such a way that we have $\langle\Phi|C_I^+ = 0 = C_I|\Phi\rangle$, $\forall I \neq 0$, with $C_0^+ \equiv 1$.

For spin systems an appropriate choice for the CCM model state $|\Phi\rangle$ is often a classical spin state,[3] in which the most general situation is one in which each spin can point in an arbitrary direction. After a local canonical transformation on each site to rotate each spin to the down ($-z$) direction, we have

$$|\Phi\rangle = |\cdots\downarrow\downarrow\downarrow\cdots\rangle; \quad C_I^+ = s_r^+,\ s_r^+ s_{r'}^+,\ s_r^+ s_{r'}^+ s_{r''}^+, \cdots, \tag{1}$$

(where the indices $r, r', r'', \ldots$ denote any lattice site) respectively, for the model state and the multispin creation operators which consist of spin-raising operators only.

The CCM parameterizations of the ket and bra ground states are given by

$$|\Psi\rangle = e^S|\Phi\rangle, \quad S = \sum_{I\neq 0} \mathcal{S}_I C_I^+, \qquad \langle\tilde{\Psi}| = \langle\Phi|\tilde{S}e^{-S}, \quad \tilde{S} = 1 + \sum_{I\neq 0} \tilde{\mathcal{S}}_I C_I. \tag{2}$$

To find the ket-state and bra-state correlation coefficients $\mathcal{S}_I$ and $\tilde{\mathcal{S}}_I$, respectively, we require that the expectation value $\bar{H} = \langle\tilde{\Psi}|H|\Psi\rangle$ be stationary with respect to variations in the bra-state and ket-state correlation coefficients. Hence, for example, the CCM ket-state equations are given by

$$\langle\Phi|C_I e^{-S} H e^S|\Phi\rangle = 0, \quad \forall I \neq 0. \tag{3}$$

The problem of determining the CCM equations now becomes a *pattern-matching exercise* of the $\{C_I\}$ to the terms in $e^{-S}He^S$ in Eq. (3). The excited states may also be parametrized in a similar fashion (and see Refs. [1, 3] for details).

We now use the lattice symmetries in order to find all *different* possible configurations with respect to the point and space group symmetries (of both the lattice and Hamiltonian) with up to n spins flips spanning a range of no more than n adjacent lattice sites (LSUBn approximation) and these are referred to as the *fundamental* configurations or clusters. Details of the CCM computational algorithm for spin quantum number $s = 1/2$ are given in Refs. [3, 9].

3. A Typical Spin-Lattice Model Exhibiting Various Phases

We consider a spin-half Heisenberg model on a square lattice with two kinds of nearest-neighbour bonds J and J', as shown in Fig. 1,

$$H = J \sum_{\langle ij\rangle_1} \mathbf{s}_i \cdot \mathbf{s}_j + J' \sum_{\langle ij\rangle_2} \mathbf{s}_i \cdot \mathbf{s}_j. \tag{4}$$

The expressions $\langle ij\rangle_1$ and $\langle ij\rangle_2$ indicate nearest-neighbour bonds arranged in a regular zigzag pattern, as shown in Fig. 1 by the dotted and solid lines, respectively. Each square-lattice plaquette consists of three J bonds and one J' bond. If J' and J have different signs then the plaquettes are frustrated, whereas competition without frustration is realized for antiferromagnetic bonds $J' > J > 0$. A model with such a zigzag pattion of bonds has been treated previously by a variety of methods.[10,11,15,16]

The case of antiferromagnetic J' bonds with $J' > J > 0$ resembles the situation in bilayer systems and in the depleted square-lattice antiferromagnet CaV_4O_9, in which the competition between two different antiferromagnetic bonds leads to a phase transition from antiferromagnetic long-range order to quantum disorder with

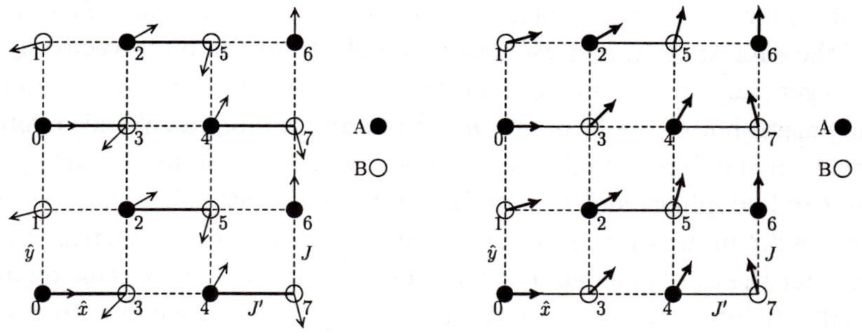

Fig. 1. Illustration of the model (J-bonds correspond to dotted lines and J'-bonds to solid lines) and of the classical spiral state for antiferromagnetic $J = +1$ and ferromagnetic $J' < -1/3$ (left graph) and vice versa ($J = -1$, $J' > 1/3$) (right graph). As discussed in the text, both spiral states can be transformed into each other by reversing all of the spins on the B sublattice.

a finite gap. It is seen in this article that the transition point obtained for the model of Eq. (4) is quite close to that obtained for the bilayer model.[17]

For special values of J and J' the model represents (i) the spin-half Heisenberg antiferromagnet (ferromagnet) on the square lattice for $J = J' = 1$ ($J = J' = -1$), (ii) the spin-half Heisenberg antiferromagnet (ferromagnet) on the honeycomb lattice for $J = 1$, $J' = 0$ ($J = -1$, $J' = 0$), and (iii) the spin-one Heisenberg antiferromagnet on the triangular lattice for $J = 1$, $J' = -\infty$.

3.1. *The classical ground state*

For $J = 1$ and $J' > -1/3$ the Néel state is the classical ground state of the Hamiltonian of Eq. (4). At $J'_c = -1/3$ there is classically a second-order phase transition to a ground state of helical nature (see Fig. 1), with a characteristic pitch angle $\Phi = \pm|\Phi_{\rm cl}|$ given by

$$|\Phi_{\rm cl}| = \begin{cases} 0 & J' > -\frac{1}{3} \\ \arccos\left(\frac{1}{2}\sqrt{1 - \frac{1}{J'}}\right) & J' \leq -\frac{1}{3}, \end{cases} \tag{5}$$

where the different signs correspond to the two chiralities[18] of this helical state. Note that for $\Phi = 0$ this is just the Néel state. More generally, the pitch angle varies with J' from $|\Phi_{\rm cl}| = 0$ for $J' > -1/3$ to $|\Phi_{\rm cl}| = \pi/3$ for $J' = -\infty$. Note that $|\Phi_{\rm cl}| = \pi/3$ (realized at $J' = -\infty$) corresponds to the ground state of the spin-1 triangular lattice. For the $J = 1$ case we describe the directions of the spins $\mathbf{s}_A$ and $\mathbf{s}_B$, belonging to the A and B sublattices respectively, for the classical helical state with a characteristic angle Φ as follows[10] (and see Fig. 1),

$$\begin{aligned} \mathbf{s}_A(\mathbf{R}) &= \hat{\mathbf{u}} \cos \mathbf{Q} \cdot \mathbf{R} + \hat{\mathbf{v}} \sin \mathbf{Q} \cdot \mathbf{R}, \\ \mathbf{s}_B(\mathbf{R} + \hat{x}) &= \hat{\mathbf{u}} \cos(\mathbf{Q} \cdot \mathbf{R} + \pi + 3\Phi) + \hat{\mathbf{v}} \sin(\mathbf{Q} \cdot \mathbf{R} + \pi + 3\Phi), \end{aligned} \tag{6}$$

where $\hat{\mathbf{u}}$ and $\hat{\mathbf{v}}$ are perpendicular unit vectors in the spin space, $\mathbf{R}$ runs over the sites of the sublattice A, and we have $\mathbf{Q} = (2\Phi, 0)$ for the pitch vector $\mathbf{Q}$. We note that this general helical state does not have a periodicity in the x-direction because Φ is in general not of the form $m\pi/n$ with m and n integral. We also note that we have only three different angles between nearest-neighbour spins, namely $\pm(\pi - \Phi)$ for the $J = 1$ couplings and $\pi - 3\Phi$ for the coupling with J'.

The maximum frustration is in the region around $J' \approx -1$. Bearing in mind the situation for the J_1–J_2 model, one might expect that for the extreme quantum case (spin-half) quantum fluctuations might be able to open a window to a spin-liquid phase for a finite range of parameters around this region of maximum frustration. On the other hand, for strong antiferromagnetic bonds ($J' \gg 1$) there is, of course, no indication in the classical model for the breakdown of the Néel order.

By contrast with the quantum case, the classical model with $J = -1$ can be simply transformed into the model with $J = +1$ considered above by the simultaneous substitutions $J \to -J$, $J' \to -J'$, $\mathbf{s}_{i \in B} \to -\mathbf{s}_{i \in B}$. Hence, the physics for the $J = -1$ case in the classical model is essentially the same as for the $J = +1$ case (and see Fig. 1).

The resulting number of CCM LSUBn configurations for $n \leq 8$ are shown in Table 1, using both the Néel and spiral states as model states.

Table 1. Number of fundamental ground-state configurations of the LSUBn approximation for the Hamiltonian of Eq. (4), using a Néel state ($\Phi = 0$) and a helical state ($\Phi \neq 0$) for the CCM model state, and the number of fundamental excited-state configurations using the Néel model state only.

LSUBn	ground state: $\Phi = 0$	$\Phi \neq 0$	excited state: $\Phi = 0$
2	3	5	1
4	22	76	16
6	267	1638	331
8	4986	42160	7863

4. Results

4.1. *Antiferromagnet: Néel versus dimer phases*

In order to set the scene we start with a simple mean-field (MF) like description of the order-disorder transition. The corresponding uncorrelated MF state for Néel longe-range order (LRO) is the Néel state $|\phi_{\mathrm{MF}_1}\rangle = |\uparrow\downarrow\uparrow \ldots\rangle$, and for the dimerized singlet state it is the rotationally-invariant product state of local pair singlets $|\phi_{\mathrm{MF}_2}\rangle = \prod_{i \in A}[|\uparrow_i\downarrow_{i+\hat{x}}\rangle - |\downarrow_i\uparrow_{i+\hat{x}}\rangle]/\sqrt{2}$ where i and $i + \hat{x}$ correspond to those sites which cover the J' bonds. In order to describe the transition between both states we consider an uncorrelated product state interpolating between $|\phi_{\mathrm{MF}_1}\rangle$ and

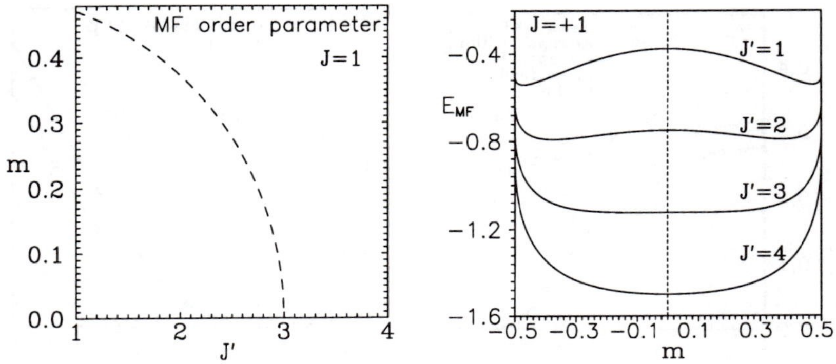

Fig. 2. Sublattice magnetization versus J' (for $J = 1$) (left graph), and energy versus sublattice magnetization (right graph), using a mean field approach as in Eq. (7).

$|\phi_{\mathrm{MF}_2}\rangle$, of the form[10,19]

$$|\Psi_{\mathrm{MF}}(t)\rangle = \prod_{i\in A} \frac{1}{\sqrt{1+t^2}} \left[|\uparrow_i\downarrow_{i+\hat{x}}\rangle - t|\downarrow_i\uparrow_{i+\hat{x}}\rangle \right]. \tag{7}$$

We have $|\Psi_{\mathrm{MF}}(t=0)\rangle = |\phi_{\mathrm{MF}_1}\rangle$ and $|\Psi_{\mathrm{MF}}(t=1)\rangle = |\phi_{\mathrm{MF}_2}\rangle$. We minimize $\langle\Psi_{\mathrm{MF}}|H|\Psi_{\mathrm{MF}}\rangle$ with respect to t and obtain

$$\frac{E_{\mathrm{MF}}}{N} = \frac{\langle\Psi_{\mathrm{MF}}|H|\Psi_{\mathrm{MF}}\rangle}{N} = \begin{cases} -\frac{3J'}{8} - \frac{1}{24J}(3J-J')^2 & J' \leq 3J \\ -\frac{3J'}{8} & J' > 3J, \end{cases} \tag{8}$$

for the energy per site. For the sublattice magnetization, $m \equiv \langle\Psi_{\mathrm{MF}}|s^z_{i\in A}|\Psi_{\mathrm{MF}}\rangle$, we get $m = \sqrt{(3J-J')(3J+J')}/(6J)$ for $J' \leq 3J$, and $m = 0$ otherwise. Note that m vanishes at a critical point $J'_c = 3J$, and that the critical index is the MF index 1/2. Equation (8) may be rewritten in terms of m as $E_{\mathrm{MF}}/N = -\frac{1}{8}J' - \frac{1}{4}J'\sqrt{1-4m^2} - \frac{3}{2}Jm^2$, and Fig. 2 illustrates that the dependence of E_{MF} on m corresponds to a typical scenario of a second-order transition (see Ref. [12]). We can expand E_{MF} up to fourth order in m near the critical point and find a Landau-type expression, given by $E_{\mathrm{MF}}/N = -\frac{3}{8}J' + \frac{1}{2}(J'-3J)\,m^2 + \frac{1}{2}J'm^4$. However, as discussed elsewhere for a similar magnetic model for CaV_4O_9,[20] MF theory is not expected to describe the critical behaviour correctly.

Let us therefore now apply a high-order CCM approach (for details see Refs. [3, 10]) to this model. We set the classical collinear Néel state to be the reference state $|\Phi\rangle$. We calculate the ground-state (GS) wave function, $|\Psi\rangle = e^S|\Phi\rangle$ within the LSUBn approximation scheme up to $n = 8$ and extrapolate to $n \to \infty$. The CCM results for the order parameter are shown in the left graph of Fig. 3 and they are compared to results of linear spin wave theory (SWT), the extrapolated results from exact diagonalization (ED) of N=16, 18, 20, 26, 32 sites, and the MF theory. The CCM is able to describe correctly the order-disorder transition, whereas conventional SWT cannot (and for more details concerning the SWT and ED results see Ref. [10]). The critical value predicted by extrapolation of the LSUBn results is,

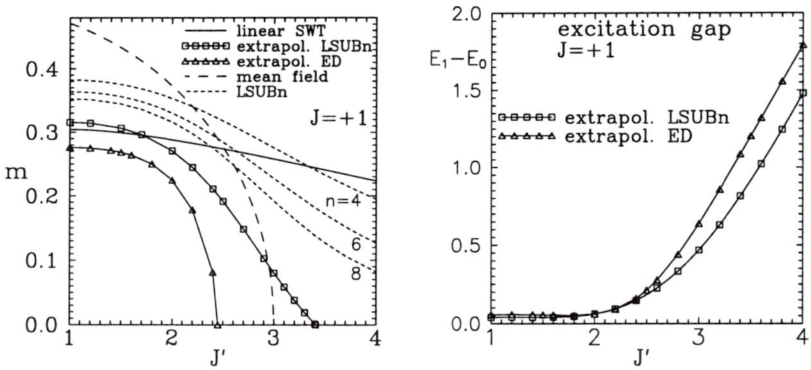

Fig. 3. Sublattice magnetization (left graph) and excitation gap (right graph) versus J'.

however, found to be slightly too large. We may also consider the inflection points of m versus J' for the LSUBn approximations. It is assumed that the true $m(J')$-curve will have a negative curvature up to the critical point. Thus we might expect that (for increasing n) the inflection point approaches the critical point. We find the corresponding inflection points at $J' = 3.1$ (n=2), $J' = 3.0$ (n=4), $J' = 2.9$ (n=6) and $J' = 2.85$ (n=8), indicating a critical value J'_c somewhere between $2.5J$ and $3J$. Notice that the estimation of $2.5 \lesssim J'_c/J \lesssim 3$ is consistent with results of series expansions and exact diagonalizations.[10,15] The breakdown of Néel LRO due to singlet formation is also accompanied by the opening of an excitation gap between the singlet GS and the first triplet excitation. This behaviour is well described by the CCM (right graph of Fig. 3) which predicts that the gap opens in the range $2J < J' < 3J$ (and notice that the non-zero gap below about $2J$ is a result of the limited accuracy of the extrapolation).

4.2. *Antiferromagnet: Néel versus spiral phases*

In Sec. 4.1 we restricted our attention to the region ($J = +1$, $J' > 0$) where we have "competition without frustration" between the bonds. The two terms within the Hamiltonian of Eq. (4) compete because the first term (J) favours Néel long-range order on the honeycomb lattice, whereas the second term (J') favours an uncorrelated dimerized state of local pair singlets. By contrast, we now turn our attention to the frustrated model (i.e., "competition with frustration") where J and J' have opposite signs. Firstly, in this subsection we consider the case $J = +1$, $J' < 0$ where the Néel and spiral states compete, and in the next subsection consider the case $J = -1$, $J' > 0$ where the ferromagnetic and spiral states compete. In both cases the ED technique for finite-size lattices is less useful because of the incommensurate nature of the classical spiral states. By contrast, the CCM considers the limit $N \to \infty$ from the outset and should have no such problem dealing with incommensurate states. In both cases we use the corresponding classical state of

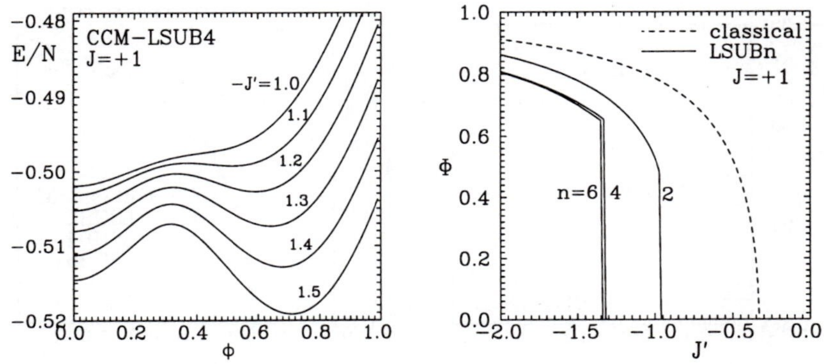

Fig. 4. Energy versus quantum pitch angle for LSUB4 (left graph) and quantum pitch angle versus J' (right graph). Note, that $\Phi = 0$ corresponds to the Néel state.

Sec. 3.1 as the CCM model state. However, since quantum fluctuations can clearly change the spiral pitch angle from its classical value, we determine the "quantum pitch angle" Φ by minimizing the energy $E_{\mathrm{LSUB}n}(\Phi)$ with respect to Φ in each order n.

Results for $E(\Phi)$ and $\Phi(J')$ are shown in Fig. 4 for the case $J = +1, J' < 0$. Several key points emerge. Firstly, it is clear that the CCM continues to yield a consistent description of the system in both these collinear and spiral phases. Secondly, the quantum Néel state remains the ground state up to much stronger frustration than in the classical case. This is in keeping with the general observation that quantum fluctuations in spin systems tend to favour collinear spin structures over non-collinear ones. Thirdly, and very strikingly, we observe that in this case the quantum fluctuations also change the order of the phase transition from second order in the classical case to first order in the quantum case.

4.3. *Ferromagnet: Ferromagnetic versus spiral phases*

We now contrast the case considered in Sec. 4.2 with the case where $J = -1$, $J' > 0$. Whereas in the classical model we again have a second-order transition between collinear and spiral phases, the quantum model in these two regimes behaves quite differently. Although the collinear antiferromagnetic phase contains strong quantum fluctuations as we saw above, the fully polarized, collinear ferromagnet contains no such fluctuations. The corresponding results for $E(\Phi)$ and $\Phi(J')$ are displayed in Fig. 5. In this case the order of the transition from the collinear (ferromagnetic) to the noncollinear (spiral) phase remains second order, as in the classical case. Furthermore, the quantum critical point remains exactly at the classical value, $J'/J = -1/3$.

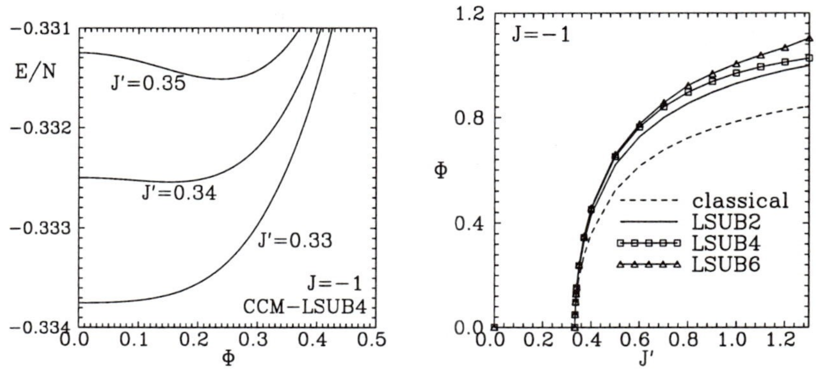

Fig. 5. Energy versus quantum pitch angle for LSUB4 (left graph) and quantum pitch angle versus J' (right graph). Note, that $\Phi = 0$ corresponds to the fully polarized ferromagnetic state.

4.4. *Comparison of the various cases*

The difference between the two cases considered in Secs. 4.2 and 4.3 is particularly illustrated by Fig. 6 which contrasts the behaviour of the order parameter, namely the on-site magnetic moment m, in the two regimes. For $J = +1$ there exists a

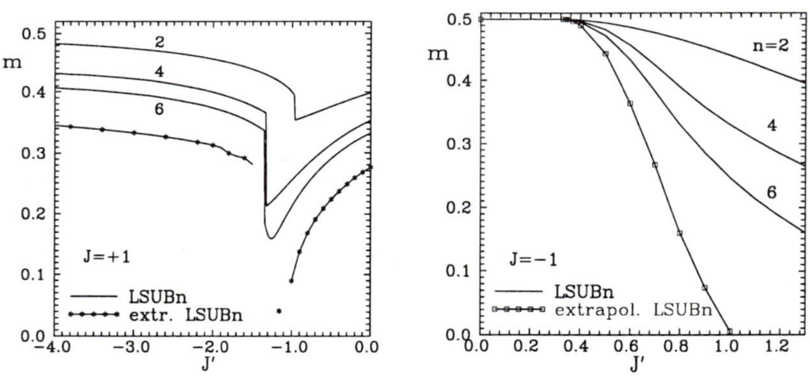

Fig. 6. On-site magnetic moment versus J' for $J = +1$ (left graph) and $J = -1$ (right graph).

discontinuity in m at every level of LSUBn approximation. This is a reflection of the behaviour observed in $E(\Phi)$ in Fig. 4, where a *local* minimum at $\Phi \neq 0$ appears for $J' \lesssim -1.1$, which for values $J' \lesssim -1.35$ becomes a *global* minimum. The extrapolation of m to the exact $n \to \infty$ limit does, however, become imprecise close to the phase transition point, and we cannot decide with certainty whether the order parameter does vanish at the transition point or, indeed, over a small finite regime around it. Figure 6 displays a minimum extrapolated value of $m \approx 0.05$ at $J'_c \approx -1.2$. We cannot exclude entirely a disordered quantum *spin-liquid phase* with $m = 0$ caused by quantum fluctuations plus frustration, but if it does exist it

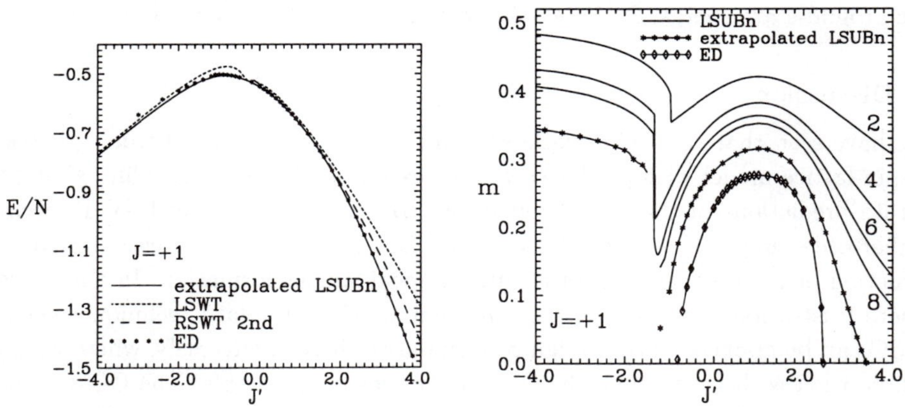

Fig. 7. Left: Ground-state energy versus J', for the case $J = +1$, for the extrapolated CCM-LSUBn approximations, in comparison with results of spin-wave theory (linear and second-order renormalized)[16], and with the extrapolated result of exact diagonalization (ED) data. Right: Ground-state magnetic order parameter versus J', for the case $J = +1$, for the CCM-LSUBn approximation. The results are compared (for the Néel region only) with the extrapolated results of exact diagonalization (ED) data. Note that both extrapolated results fit poorly in a region around $J' \approx -1$, and we therefore plot them here as isolated points (omitting the solid lines).

can clearly occur only over a very narrow regime in J' around $-1.5 \lesssim J' \lesssim -1.1$. The most likely scenario however is a single phase boundary at a value $J'_c \approx -1.35$ between spiral and Néel phases, with no separate spin-liquid phase.

By contrast, Fig. 6 shows a smooth change in m at the critical point $J'_c = 1/3$ for $J = -1$. For increasing $J' > J'_c$ the spiral magnetic order becomes weaker and finally vanishes at $J' \approx 1$. The underlying reason for this is again local singlet formation (i.e., dimerization), as discussed in Sec. 4.1. On the other hand, a much smaller critical strength J' is now needed for dimerization than in the case $J = +1$ (where $2.5 \lesssim J'_c \lesssim 3$), due to the effects of frustration which now assist in the formation of local singlets.[19]

Finally, in Fig. 7 we show both the ground-state energy $E(J')$ and the order parameter $m(J')$ in the case $J = +1$ over the regimes spanning the spiral, Néel, and dimer-like phases. We observe excellent agreement between the CCM results and the results from exact diagonalization (ED) for $J' > 0$. By contrast, spin-wave theory (SWT) calculations[16] show a significant deviation for larger $J' >> 1$. These latter are obviously poor since they lie *above* the simple variational upper bound of Eq. (8). The CCM and ED results, by comparison, lie slightly below the mean-field variational result.

We note that although the Néel state is the starting reference state for both CCM and SWT calculations, the CCM is much better able than SWT to describe the transition to the rotationally-invariant disordered state and hence to its limiting form, namely the fully dimerized state of Eq. (7) with $t = 1$. For example, even the simplest LSUB2 CCM approximation gives the correct asymptotic behaviour for the ground-state energy, $E/N \to -3J'/8$ as $J' \to \infty$, whereas SWT does not.

5. Discussion

We have seen that the CCM works extremely well over several distinct phases of a spin-lattice model with a relatively complex phase diagram, including giving accurate predictions of the phase boundaries. By contrast, standard SWT, even in higher-order implementations, fails to describe the transition caused by the destruction of Néel longe-range order due to local singlet formation. In this region, where frustration is absent, on the other hand, other standard techniques such as QMC can be applied, since no sign problem arises here. Conversely, where we have competition with frustration, the QMC is not easily applicable. The CCM, almost alone among universal methods, yields a consistent description of collinear, noncollinear, and disordered phases. It has even been shown to be able to describe the quantum first-order transition from the Néel phase to the helical phase, for which the classical model has a second-order transition.

Based on the success of the CCM to describe this and other complex magnetic lattice models, it would be interesting to extend the method both to other strongly correlated electron-lattice models and to include the effects of finite (nonzero) temperatures. In the latter regard one may then expect to be able to use the CCM to investigate quantitatively the competing effects of quantum and thermal fluctuations in a fully microscopic formalism.

References

1. R.F. Bishop, in *Microscopic Quantum Many-Body Theories and Their Applications*, edited by J. Navarro and A. Polls, Lecture Notes in Physics Vol. **510** (Springer-Verlag, Berlin, 1998), p. 1.
2. R.F. Bishop, *Theor. Chim. Acta* **80**, 95 (1991); R.J. Bartlett, *J. Phys. Chem.* **93**, 1697 (1989).
3. C. Zeng, D.J.J. Farnell, and R.F. Bishop, *J. Stat. Phys.* **90**, 327 (1998); D.J.J. Farnell, R.F. Bishop, and K.A. Gernoth, *J. Stat. Phys.* **108**, 401 (2002).
4. D.J.J. Farnell and R.F. Bishop, in *Quantum Magnetism*, edited by U. Schollwöck, J. Richter, D.J.J. Farnell, and R.F. Bishop, Lecture Notes in Physics Vol. **645** (Springer-Verlag, Berlin, 2004), p. 307.
5. R.F. Bishop, R.G. Hale, and Y. Xian, *Phys. Rev. Lett.* **73**, 3157 (1994).
6. D.J.J. Farnell, S.E. Krüger, and J.B. Parkinson, *J. Phys.: Condens. Matter* **9**, 7601 (1997).
7. R.F. Bishop, D.J.J. Farnell, and J.B. Parkinson, *Phys. Rev. B* **58**, 6394 (1998).
8. R.F. Bishop, D.J.J. Farnell, and C. Zeng, *Phys. Rev. B* **59**, 1000 (1999).
9. R.F. Bishop, D.J.J. Farnell, S.E. Krüger, J.B. Parkinson, J. Richter, and C. Zeng, *J. Phys.: Condens. Matter* **12**, 6887 (2000).
10. S.E. Krüger, J. Richter, J. Schulenburg, D.J.J. Farnell, and R.F. Bishop, *Phys. Rev. B* **61**, 14607 (2000).
11. S.E. Krüger and J. Richter, *Phys. Rev. B* **64**, 024433 (2001).
12. J. Richter, S.E. Krüger, D.J.J. Farnell, and R.F. Bishop, in *150 Years of Quantum Many-Body Theory: A Festschrift in Honour of the 65th Birthdays of John W. Clark,*

Alpo J. Kallio, Manfred L. Ristig, and S. Rosati, edited by R.F. Bishop, K.A. Gernoth, and N.R. Walet, Series on Advances in Quantum Many-Body Theory Vol. **5** (World Scientific, Singapore, 2001), p. 239.
13. D.J.J. Farnell, R.F. Bishop, and K.A. Gernoth, *Phys. Rev. B* **63**, 220402(R) (2001).
14. R.F. Bishop, Y. Xian, and C. Zeng, *Int. J. Quantum Chem.* **55**, 181 (1995).
15. R.R.P. Singh, M.P. Gelfand, and D.A. Huse, *Phys. Rev. Lett.* **61**, 2484 (1988).
16. N.B. Ivanov, S.E. Krüger, and J. Richter, *Phys. Rev. B* **53**, 2633 (1996).
17. A.W. Sandvik and D.J. Scalapino, *Phys. Rev. Lett.* **72**, 2777 (1994).
18. J. Villain, *J. Phys. C: Solid State* **10**, 4793 (1977).
19. C. Gros, W. Wenzel, and J. Richter, *Europhys. Lett.* **32**, 747 (1995).
20. M. Troyer, M. Imada, and K. Ueda, *J. Phys. Soc. Jpn.* **66**, 2957 (1997).

RECENT APPLICATIONS OF THE DMRG METHOD

KAREN HALLBERG

Centro Atómico Bariloche and Instituto Balseiro, CNEA, 8400 Bariloche, Argentina,
karen@cab.cnea.gov.ar

Received 1 December 2004

Since its inception, the DMRG method has been a very powerful tool for the calculation of physical properties of low-dimensional strongly correlated systems. It has been adapted to obtain dynamical properties and to consider finite temperature, time-dependent problems, bosonic degrees of freedom, the treatment of classical problems and non-equilibrium systems, among others. We will briefly review the method and then concentrate on its latest developments, describing some recent successful applications. In particular we will show how the dynamical DMRG can be used together with the Dynamical Mean Field Theory (DMFT) to solve the associated impurity problem in the infinite-dimensional Hubbard model. This method is used to obtain spectral properties of strongly correlated systems. With this algorithm, more complex problems having a larger number of degrees of freedom can be considered and finite-size effects can be minimized.

Keywords: Density matrix renormalization; strongly correlated electrons; low dimensional systems; dynamical mean field theory.

1. Introduction

The Density Matrix Renormalization Group was developed by S. White in 1992[1] and since then DMRG has proved to be a very powerful method for low dimensional interacting systems. Its remarkable accuracy can be seen for example in the spin-1 Heisenberg chain: for a system of hundreds of sites a precision of 10^{-10} for the ground state energy can be achieved. It has been applied to a great variety of systems and problems including, among others, spin chains and ladders, fermionic and bosonic systems, disordered models, impurities and molecules and 2D electrons in high magnetic fields. It has also been improved substantially in several directions like two (and three) dimensional (2D) classical systems, stochastic models, problems with phonons, quantum chemistry, field theory, the inclusion of temperature and the calculation of dynamical and time-dependent properties. Some calculations have also been performed in 2D quantum systems. These topics are treated in detail and in a pedagogical way in the book,[2] where the reader can find an extensive review on DMRG. Recent reviews on DMRG can also be found in 3 and 4.

When considering finite systems, the exponential growth of degrees of freedom to be considered imposes an important limitation in numerical calculations. Several methods have been introduced in order to reduce the size of the Hilbert space to be able to reach larger systems, such as Monte Carlo, renormalization group (RG) and DMRG. Each method considers a particular criterion for keeping the relevant information. For example, in RG methods, only the lowest-lying states are kept in each building block having an excellent performance for the Kondo and Anderson impurity models (Wilson NRG)[5] but a very poor outcome for strongly correlated systems like the Hubbard model.

In the following Section we will briefly describe the standard method; in Sect. 3 we will mention some of the most important applications and extensions; in Sect.4 we describe how to calculate dynamical properties and finally, in Sect. 5 we will describe how DMRG can be used within the Dynamical Mean Field theory (DMFT) to improve the performance of this latter method.

2. Basic Facts

The DMRG allows for a systematic truncation of the Hilbert space by keeping the most probable states in describing a wave function (*e.g.* the ground state) of a larger system, instead of the lowest energy states usualy kept in previous real space renormalization techniques. The method is very well described in Ref. [1] but I shall summarize it so as to unify notations. Let's define block [**B**] as a finite chain with N_B sites, having an associated Hilbert space where operators are defined (specially the Hamiltonian in this finite chain and the operators at the ends of the block, useful for linking it to other chains or added sites). We also define an added block or site as [**a**] (see Fig. 1)

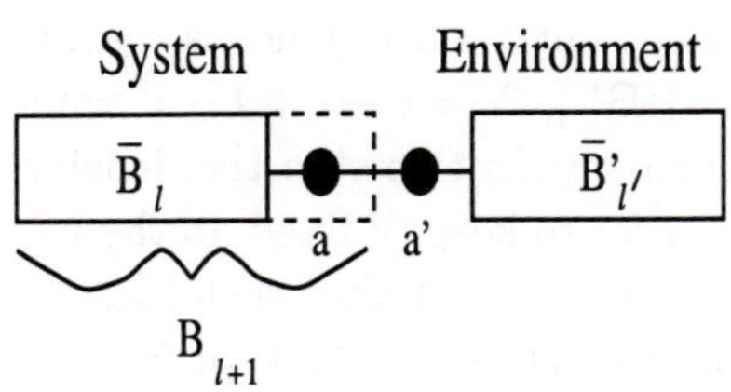

Fig. 1. A scheme of the superblock (universe) configuration for the DMRG algorithm.[1]

Below we describe a general iteration of the method:

i) Define the Hamiltonian for the superblock formed by putting together two blocks [**B**] and two added sites in such a way: [**B a B' a'**] (the primes are only to indicate additional blocks, but the blocks have the same structure as the non-primed ones; however, this is not necessary and smaller blocks can be added). In general, blocks [**B**] and [**B'**] come from the previous iteration. The total Hilbert space of this superblock is a direct product of the individual spaces corresponding to each block and the added sites. Here periodic boundary conditions can be attached between [**a'**] and [**B**] (the ends of the superblock). Diagonalize the Hamiltonian to obtain the ground state $|\psi_0\rangle$ (other states could also be kept, such as the first excited ones: they are called target states).

ii) Construct the density matrix $\rho_{ii'} = \sum_j \psi_{0,ij}\psi_{0,i'j}$ on block [**B a**] where $\psi_{0,ij} = \langle i \otimes j|\psi_0\rangle$, the states $|i\rangle$ belonging to the Hilbert space of the block [**B a**] and the states $|j\rangle$ to the block [**B' a'**]. The density matrix considers the part [**B a**] as a system and the rest, (*i.e.* [**B' a'**]), as a statistical bath. The eigenstates of ρ with the highest eigenvalues correspond to the most probable states (or equivalently the states with higher weight) of block [**B a**] in the ground state defined in the whole superblock. These states are kept up to a certain cutoff, keeping a total of m states per block. The density matrix eigenvalues sum up to unity and the truncation error, defined as the sum of the density matrix eigenvalues corresponding to discarded eigenvectors, gives a qualitative indication of the accuracy of the calculation.

iii) These states form a new reduced basis to which all the operators have to be changed and the block [**B a**] is renamed as block [**B**].

iv) A new block [**a**] is added (one site in our case) and the new superblock [**B a B' a'**] is formed as the direct product of the states of all the blocks.

v) This iteration continues until the desired length is achieved. At each step the length is $2N_B+2$ (if [**a**] consists of one site).

When more than one target state is used, *i.e* more than one state is wished to be well described, the density matrix is defined as $\rho_{ii'} = \sum_l p_l \sum_j \phi_{l,ij}\phi_{l,i'j}$, where p_l defines the probability of finding the system in the target state $|\phi_l\rangle$ (not necesseraly eigenstates of the Hamiltonian).

The method described above is usually called the *infinite-system algorithm* since the system size increases at each iteration. There is a way to increase precision at each length N called the *finite-system algorithm.* It consists of fixing the lattice size and zipping a couple of times until convergence is reached. In this case and for the block configuration [**B a a' B'**], $N = l+1+1+l'$ where l and l' are the number of sites in B and B' respectively. In this step the density matrix is used to project onto the left $l+1$ sites. In order to keep N fixed, in the next block configuration, the right block B' should be defined in $l-1$ sites such that $N = (l+1)+1+1+(l-1)'$. The operators in this smaller block should be kept from previous iterations (in some cases from the iterations for the system size with $N-2)^2$.

To calculate static properties like correlation functions one must keep the relevant operators at each step and perform the corresponding basis change and reduction, in a similar manner as done with the Hamiltonian in each block[1]. The energy and measurements are calculated in the superblock. In Ref. 7 an interpretation of the correlation functions of systems at criticality is given in terms of wave function entanglement, conjecturing a modification of DMRG for these cases that preserves the entanglement.

In connection with the above, recent work[8,9] has focused on the quantum information perspective of DMRG by calculating the von Neumann entropy of entanglement. Within this frame it can be shown that DMRG, in its present form, is not suited for calculations in dimensions higher than one. Although calculations in 2D have been performed, the DMRG does not have the potentiality it has for 1D systems.

An interesting analysis on DMRG accuracy is done in Ref. 10. Fixed points of the DMRG and their relation to matrix product wave functions were studied in 11 and an analytic formulation combining the block renormalization group with variational and Fokker-Planck methods in 12. The connection of the method with quantum groups and conformal field theory is treated in 13. There are also interesting connections between the density matrix spectra and integrable models[14] via corner transfer matrices.

3. Applications and Extensions to DMRG

The number of papers using DMRG has grown enormously since its development. For example, since 1998 there have been around 80 papers using DMRG published a year. There have also been several improvements to the method and it is now used in areas that are very different to the original strongly correlated electron system field. We will only mention very briefly the main applications here and refer the reader to the reviews on the subject were most of the references are given as well as some details of the calculations.[3,4]

One of the first calculations that showed the potential of the method was obtained by White and Huse[15] when calculating the spin gap in a $S = 1$ Heisenberg chain and very precise spin correlation functions and excitation energies. Several calculations were devoted to spin chains and ladders, with different values of total spin obtaining very accurate results for the excitation spectra, thermodynamics and scaling behaviour.[16]

The application of DMRG to fermionic systems has been very extensive and interesting results have been obtained for one-dimensional Hubbard and t-J models and their generalizations. Bosonic degrees of freedom like phonons were introduced and applied to calculate properties of the Holstein model. Pure bosonic systems have also been calculated such as the Bose-Hubbard model. It has been generalized to 1D random and disordered systems, and applied to the random antiferromagnetic and ferromagnetic Heisenberg, including quasiperiodic exchange modulation and other systems.

A first step towards higher dimensions was the application to ladder systems (spin and fermionic ones) to study different properties like magnetization plateaus. An interesting review on the applications to some exact and analytical techniques for quantum magnetism in low dimension, including DMRG, is presented in 17. Quite large two-dimensional systems could be reached considering several leg ladders or different sweeping algorithms in square lattices, but there's no reliable and systematic implementation of the method in 2D yet and its performance is still much poorer than in 1D. Recently, there has been an interesting approach to 2D quantum systems based on matrix product states.[8] Long-range Coulomb interactions leading to several phases have been found using DMRG for 2D electrons in high magnetic fields considering different Landau levels.

Impurity problems have been studied for example in one- and two-impurity Kondo systems, in spin chains and in Luttinger Liquids. There have also been applications to Kondo and Anderson lattices and ladders and models for manganites, the so-called Ferromagnetic Kondo Lattice model (see e.g. 18, 19 for recent results).

There have been several improvements to the DMRG method like the inclusion of symmetries such as total spin, parity and momentum representation.[20] Other recent applications have been in nuclear shell model calculations where a two level pairing model has been considered[21] and in the study of ultrasmall superconducting grains, in this case, using the particle (hole) states around the Fermi level as the system (environment) block.[22,23] It has also been used in problems in high energy physics in toy models for QCD[24] and field theory models.

The DMRG has been very successfully extended to study classical systems. For a detailed description we refer the reader to Ref. 25. Since 1D quantum systems are related to 2D classical systems,[26] it is natural to adapt DMRG to the classical 2D case. This method is based on the renormalization group transformation for the transfer matrix T (TMRG).[27] It is a variational method that maximizes the partition function using a limited number of degrees of freedom, where the variational state is written as a product of local matrices.[11] For 2D classical systems, this algorithm is superior to the classical Monte Carlo method in accuracy, speed and in the possibility of treating much larger systems. It was first applied to the Ising and Potts models and non-hermitian problems in equilibrium and non-equilibrium physics.

The application of the DMRG method to classical systems opens the possibility to study 1D quantum systems at non zero temperature, by using the Trotter-Suzuki method [28,29,30,31,32] where the results are precise at high temperatures. In this case the system is infinite and the finiteness is in the level of the Trotter approximation. It has been applied to calculate thermodynamic properties for spin chains and ladders, Kondo insulator, Kondo impurities and t-J models.

A very promising extension of DMRG concerns molecules and its utilization in Quantum Chemistry. It has been used to calculate properties in long organic molecules, large Fe molecules, models for reptons and excitations in dendrimer molecules. Recent attempts to apply DMRG to the *ab initio* calculation of electronic states in molecules have been successful.[33,34,35,36]

Another recent and important development concerns real-time calculations to study for example the evolution of wave functions or the calculation of transport properties in low-dimensional systems (the so-called time dependent DMRG). Early calculations based on the integration of the Schrödinger equation were performed in Ref. 37, 38 and applied to the calculation of transport properties through a junction between two Luttinger liquids.[39] A very recent improvement based on the Trotter-Suzuki expansion of the evolution operator[40] was done by White and Feiguin,[41] where they obtain vary accurate results and even obtain the frequency dynamics by Fourier transforming the real-time results.

4. DMRG for Dynamical Properties

DMRG can also be used to calculate dynamical properties of low-dimensional systems, useful to interpret experimental results from, for example, nuclear magnetic resonance (NMR), neutron scattering, optical absorption, photoemission, among others. There have been two main approaches to the dynamics, the Lanczos[42,43] and the correction vector techniques.[43,44,45] The first gives complete information of the whole excitation spectrum at the expense of less accuracy for large systems, specially at high energies. The latter, instead, focuses on particular energy values and give more precise information, but numerically much more expensive.

4.1. *Lanczos technique*

Suppose one wants to calculate the following dynamical correlation function at $T = 0$, $C_A(t-t') = \langle\psi_0|A^\dagger(t)A(t')|\psi_0\rangle$, where $A^\dagger$ is the Hermitean conjugate of the operator A, $A(t)$ is the Heisenberg representation of A, and $|\psi_0\rangle$ is the ground state of the system. Its Fourier transform is:

$$C_A(\omega) = \sum_n |\langle\psi_n|A|\psi_0\rangle|^2 \,\delta(\omega - (E_n - E_0)), \tag{1}$$

where the summation is taken over all the eigenstates $|\psi_n\rangle$ of the Hamiltonian H with energy E_n, and E_0 is the ground state energy.

Defining the Green's function $G_A(z) = \langle\psi_0|A^\dagger(z-H)^{-1}A|\psi_0\rangle$, the correlation function $C_A(\omega)$ can be obtained as

$$C_A(\omega) = -\frac{1}{\pi}\lim_{\eta\to 0^+} \mathrm{Im}\, G_A(\omega + i\eta + E_0). \tag{2}$$

The function G_A can be written in the form of a continued fraction:

$$G_A(z) = \cfrac{\langle\psi_0|A^\dagger A|\psi_0\rangle}{z - a_0 - \cfrac{b_1^2}{z - a_1 - \cfrac{b_2^2}{z - \ldots}}} \tag{3}$$

The coefficients a_n and b_n can be obtained using the following recursion equations:[46,47] $|f_{n+1}\rangle = H|f_n\rangle - a_n|f_n\rangle - b_n^2|f_{n-1}\rangle$, where $|f_0\rangle = A|\psi_0\rangle$, $a_n = \langle f_n|H|f_n\rangle/\langle f_n|f_n\rangle$ and $b_n^2 = \langle f_n|f_n\rangle/\langle f_{n-1}|f_{n-1}\rangle$, with $b_0 = 0$.

For finite systems the Green's function $G_A(z)$ has a finite number of poles so only a certain number of coefficients a_n and b_n have to be calculated. An important requirement is that the reduced Hilbert space should also describe with great precision the relevant excited states $|\psi_n\rangle$. This is achieved by choosing the appropriate target states. For most systems it is enough to consider as target states the ground state $|\psi_0\rangle$ and the first few $|f_n\rangle$ with $n = 0, 1...$ and $|f_0\rangle = A|\psi_0\rangle$ as described above. In doing so, states in the reduced Hilbert space relevant to the excited states connected to the ground state via the operator of interest A are included.

This method has been successfully applied to a number of problems, like spin 1/2 and 3/2[42,16] and spin 1 chains[43], the spin-boson model[48], the Holstein model[49] and spin-orbital chains in external fields[50]. It was also applied to extract spin-chain dispersion relations,[51] dynamics of spin ladders,[52] spectral functions in the infinite-U Hubbard model[53] and optical response in 1D Mott insulators.[54] In Section 4 we will describe its application as the impurity solver within the Dynamical Mean Field Theory method (DMFT).

4.2. *Correction vector dynamics*

This method focuses on a particular energy or energy window, allowing for a more precise description in that range and the possibility of calculating spectra for higher energies. Instead of using the tridiagonalization of the Hamiltonian, but in a similar spirit regarding the important target states to be kept, the spectrum can be calculated for a given $z = w + i\eta$ by using a correction vector (related to the operator A that can depend on momentum q).

Following the Green's function given above, the (complex) correction vector $|x(z)\rangle$ can be defined as:

$$|x(z)\rangle = \frac{1}{z - H} A|\psi_0\rangle \tag{4}$$

so the Green's function can be calculated as $G(z) = \langle\psi_0|A^\dagger|x(z)\rangle$.

Separating the correction vector in real and imaginary parts $|x(z)\rangle = |x^r(z)\rangle + i|x^i(z)\rangle$ we obtain

$$\begin{aligned} ((H - w)^2 + \eta^2)|x^i(z)\rangle &= -\eta A|\psi_0\rangle \\ |x^r(z)\rangle &= \frac{1}{\eta}(w - H)|x^i(z)\rangle \end{aligned} \tag{5}$$

The former equation is solved using the conjugate gradient method. In order to keep the information of the excitations at this particular energy the following states are targeted in the DMRG iterations: The ground state $|\psi_0\rangle$, the first Lanczos vector $A|\psi_0\rangle$ and the correction vector $|x(z)\rangle$. Even though only a certain energy is focused on, DMRG gives the correct excitations for an energy range surrounding this particular point so that by running several times for nearby frequencies, an approximate spectrum can be obtained for a wider region.[43]

A variational formulation of the correction vector technique has been developed in 45. It has been successfully applied to calculate the optical conductivity of Mott insulators.[55,56,57]

The correction vector model has also been applied to determine the nonlinear optical coefficients of Hubbard chains and related models,[58] to calculate ac conductivity of the Bose-Hubbard model[59] and the single-impurity Anderson model.[60]

5. Dynamical Mean Field Theory using DMRG

The Dynamical Mean Field Theory (DMFT) has become one of the basic methods to calculate realistic electronic band structure in strongly correlated systems.[61] At the heart of the DMFT method is the solution of an associated quantum impurity model where the environment of the impurity has to be determined self-consistently. Therefore the ability to obtain reliable DMFT solutions of lattice model Hamiltonians relies directly on the ability to solve quantum impurity models. Among the *a priori* exact numerical algorithms available we count the Hirsch-Fye Quantum Monte Carlo[62,63] method and Wilson's Numerical Renormalization Group (NRG).[64,65,66] While the former, a finite-temperature method, is very stable and accurate at the Matsubara frequencies, its main drawback is the access to real frequency quantities for the calculation of spectral functions which requires less controlled techniques for the analytic continuation of the Green functions. The second method can be formulated both at T=0 and finite (small) T and provides extremely accurate results at very small frequencies, at the expense of a less accurate description of the high energy features.

In order to overcome the difficulties encountered by these other methods, we will show that DMRG can be used very reliably to solve the related impurity problem within DMFT.[67] By using the DMRG to solve the related impurity problem, no *a priori* approximations are made and the method provides equally reliable solutions for both gapless and gapfull phases. More significantly, it provides accurate estimates for the distributions of spectral intensities of high frequency features such as the Hubbard bands, that are of main relevance for analysis of x-ray photoemission and optical conductivity experiments.

We will now very briefly describe the method applied to the Mott thansition in the Hubbard model. The Hamiltonian of the Hubbard model is defined by

$$H = t \sum_{\langle i,j \rangle, \sigma} c^{\dagger}_{i,\sigma} c_{j,\sigma} + U \sum_{i} n_{i,\downarrow} n_{i,\uparrow} \tag{6}$$

Applying DMFT to this model leads to a mapping of the original lattice model onto an associated quantum impurity problem in a self-consistent bath. In the particular case of the Hubbard model, the associated impurity problem is the single impurity Anderson model (SIAM), where the hybridization function $\Delta(\omega)$, which in the usual SIAM is a flat density of states of the conduction electrons, is now to be determined self-consistently. More precisely, for the Hamiltonian (6) defined on a Bethe lattice of coordination d, one takes the limit of large d and exactly maps the model onto a SIAM impurity problem with the requirement that $\Delta(\omega) = t^2 G(\omega)$, where $G(\omega)$ is the impurity Green's function. At the self-consistent point $G(\omega)$ coincides with the *local* Green's function of the original lattice model.[68] We take the half-bandwidth of the non interacting model as unit of energy, $t = 1/2$.

The Green's function of the impurity problem is an important quantity in this algorithm: $\mathrm{G}_0(\omega) = 1/(\omega + \mu - \Delta(\omega)) = 1/(\omega + \mu - t^2 G(\omega))$. Thus, to implement the new algorithm we shall consider[69,70] a general representation of the hybridization function in terms of continued fractions that define a *parametrization* of $\Delta(\omega)$ in terms of a set of real and positive coefficients. Since it is essentially a Green's function, $\Delta(z)$ can be decomposed into "particle" and "hole" contributions as $\Delta(z) = \Delta^>(z) + \Delta^<(z)$ with $\Delta^>(z) = t^2\langle gs|c\frac{1}{z-(H-E_0)}c^\dagger|gs\rangle$ and $\Delta^<(z) = t^2\langle gs|c^\dagger\frac{1}{z+(H-E_0)}c|gs\rangle$ for a given Hamiltonian, H with ground-state energy E_0. By standard Lanczos technique, H can be in principle tri-diagonalized and the functions $\Delta^>(z)$ and $\Delta^<(z)$ can be expressed in terms of respective continued fractions.[42] As first implemented in Ref. 69, 70, each continued fraction can be represented by a chain of auxiliary atomic sites whose energies and hopping amplitudes are given by the continued fraction diagonal and off-diagonal coefficients respectively.

As a result of the self-consistency condition, the two chains representing the hybridization, are "attached" to the right and left of an atomic site to obtain a new SIAM Hamiltonian, H. In fact $G_0(z)$ constitutes the local Green's function of the site plus chain system. The algorithm in Ref. 69, 70, basically consists in switching on the local Coulomb interaction at the impurity site of the SIAM Hamiltonian and use the Lanczos technique to re-obtain $\Delta(z)$, iterating the procedure until the set of continued fractions coefficients converges. By using the DMRG, the number of auxiliary sites that can be used in the hybridization chains is much larger than in the exact diagonalization scheme, leading to more accurate results.[71] An alternative way of using DMRG to solve the impurity problem, which does not rely on the continued fraction expansion was developed in 72.

The SIAM Hamiltonian therefore reads

$$H = \sum_{\sigma,\alpha=-N_C;\alpha\neq 0}^{N_C} a_\alpha c^\dagger_{\alpha\sigma}c_{\alpha\sigma} + \sum_{\sigma,\alpha=-(N_C-1);\alpha\neq 0,-1}^{N_C-1} b_\alpha(c^\dagger_{\alpha\sigma}c_{\alpha+1\sigma} + h.c.)$$
$$+ \sum_{\sigma,\alpha=\pm 1} b_0(c^\dagger_\sigma c_{\alpha\sigma} + h.c.) + U(n_\uparrow - \frac{1}{2})(n_\downarrow - \frac{1}{2}) \tag{7}$$

with c_σ being the destruction operator at the impurity site, and $c_{\alpha\sigma}$ being the destruction operator at the α site of the hybridization chain of $2N_C$ sites. The set of parameters $\{a_\alpha, b_\alpha\}$ are directly obtained from the coefficients of the continued fraction representations of $\Delta(z)$ by the procedure just described.

In Fig.2 we show the DMFT+DMRG results (solid lines) for the density of states (DOS) for several values of increasing interaction U. The results are compared to the Iterated Perturbation Theory (IPT) results (dashed lines).[73,74]

Results for the imaginary part of the Green's funtions on the Matsubara frequencies match the precise Monte Carlo solutions at low temperatures. We also obtained accurate values for the two distinct critical values of the interaction $U_{c1} == 2.39 \pm 0.02$ and $U_{c2} = 3.0 \pm 0.2$. To this end, reaching larger system sizes turned out to be important in order to perform proper extrapolations and overcome finite-size effects.

6. Conclusions

In summary, the DMRG method has proven to be a very reliable and versatile numerical method that can be applied to a broad spectrum of problems in physics. In

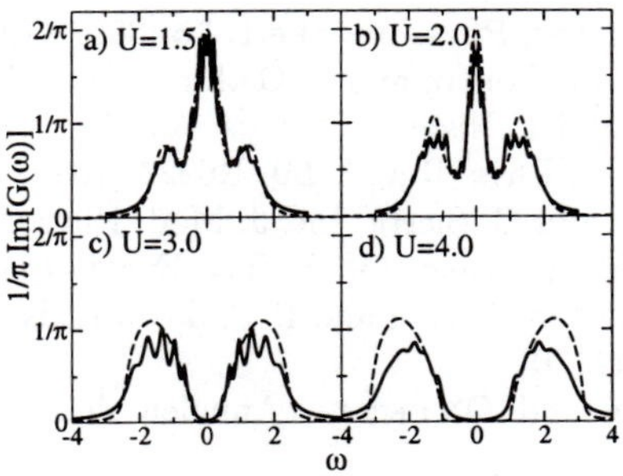

Fig. 2. Density of states ($\frac{1}{\pi}\mathrm{Im}G(\omega)$) for the half-filled Hubbard Model. We also show the IPT results (dashed lines) (see text).

spite of having been developed twelve years ago, it's still evolving and it presents very interesting potential new applications. One of them was presented here for solving the DMFT equations of strongly correlated models exploiting the DMRG methodology. Large systems can be considered and accurate values of the critical interactions are obtained in agreement with NRG predictions allowing for a non-trivial test of the accuracy of this method. In contrast with NRG, however, this new algorithm deals with all energy scales on equal footing which allowed us to find interesting substructure in the Hubbard bands of the correlated metallic state. The ability of the new algorithm to directly deal with the high energy scales is a very important feature which is relevant for the interpretation of high resolution photoemission spectroscopies.[75] In addition, with this method, realistic band-structure calculations of systems with a larger number of degrees of freedom can be handled.

References

1. S. White, Phys. Rev. Lett. **69**, 2863 (1992); S. White, Phys. Rev. B **48**, 10345 (1993)
2. *Density Matrix Renormalization*, edited by I. Peschel, X. Wang, M. Kaulke and K. Hallberg (Series: Lecture Notes in Physics, Springer, Berlin, 1999)
3. A. Schollwöck, *The density matrix renormalization group*, to appear in Rev. Mod. Phys., Jan. 2004
4. K. Hallberg, *Density Matrix Renormalization: A Review of the Method and its Applications*, in: D. Senechal, A-.M. Tremblay and C. Bourbonnais (eds.), *Theoretical Methods for Strongly Correlated Electrons*, CRM Series in Mathematical Physics, Springer (2003)
5. K. Wilson, *Rev. Mod. Phys.* **47**, 773 (1975)
6. S. White and R. Noack, Phys. Rev. Lett. **68**, 3487, (1992); R. Noack and S. White, Phys. Rev. B **47**, 9243 (1993)
7. T. Osborne and M. Nielsen, Quantum Information Processing, Volume 1, Issue 1-2, 45 (2002) (available on-line: http://www.kluweronline.com/issn/1570-0755/)
8. F. Verstraete and J. Cirac, preprint, cond-mat/0407066
9. F.Verstraete et al., Phys. Rev. Lett. **93** 227205 (2004); Latorre et al., Quant. Inf. Comp. **6**, 48 (2004); J. Gaite, preprint, quant-ph/0301120
10. Ö. Legeza and G. Fáth, Phys. Rev. B **53**, 14349 (1996); M-B Lepetit and G. Pastor, Phys. Rev. B **58**, 12691 (1998)

11. S. Östlund and S. Rommer, Phys. Rev. Lett. **75**, 3537 (1995); Phys. Rev. B **55**, 2164 (1997); M. Andersson, M. Boman and S. Östlund, Phys. Rev. B **59**, 10493 (1999); H. Takasaki, T. Hikihara and T. Nishino, J. Phys. Soc. Jpn. **68**, 1537 (1999); K. Okunishi, Y. Hieida and Y.Akutsu, Phys. Rev. E **59**, R6227 (1999)
12. M. A. Martín-Delgado and G. Sierra, Int. J. Mod. Phys. A, **11**, 3145 (1996)
13. G. Sierra and M. A. Martín-Delgado, in "The Exact Renormalization Group " by a Krasnitz, R Potting, Y A Kubyshin and P. S. de Sa (Eds.) World Scientific Pub. Co., (1999), (cond-mat/9811170)
14. I. Peschel, M. Kaulke and Ö. Legeza, Annalen der Physik **8**, 153 (1999),(cond-mat/9810174)
15. S. R. White and D. Huse, Phys. Rev. B **48**, 3844 (1993)
16. K. Hallberg, P. Horsch and G. Martínez, Phys. Rev. B, **52**, R719 (1995); K. Hallberg, X. Q. G. Wang, P. Horsch and A. Moreo, Phys. Rev. Lett. **76**, 4955 (1996).
17. S. K. Pati, S. Ramasesha and D. Sen, in *Magnetism: Molecules to Materials IV*, eds. J.S. Miller and M. Drillon (Wiley-VCH Weinheim, 2002), Chapter 4, (cond-mat/0106621)
18. D. García, K. Hallberg, C. Batista, M. Avignon and B. Alascio, Phys. Rev Lett., **85**, 3720 (2000); D. García, K. Hallberg, B. Alascio and M. Avignon, Physical Rev. Lett., **93** num. 17, 177204 (2004)
19. I.P. McCulloch *et al.*, Phys. Rev. B, **65**,52410 (2002)
20. T. Xiang, Phys. Rev. B **53**, R10445 (1996)
21. J. Dukelsky and G. Dussel, Phys. Rev. B **59**, R3005 (1999) and references therein.
22. J. Dukelsky and G. Sierra, Phys. Rev. Lett. **83**, 172 (1999)
23. J. Dukelsky and S. Pittel, Rep. Prog. Phys. **67**, 513 (2004)
24. M. A. Martín-Delgado and G. Sierra, Phys. Rev. Lett. **83**, 1514 (1999)
25. T. Nishino, J. Phys. Soc. Jpn. **64**, 3598 (1995); see also T. Nishino in Ref. 2, Chap. 5(I); T. Nishino and K. Okunishi in *Strongly Correlated Magnetic and Superconducting Systems*, Ed. G. Sierra and M. A. Martín-Delgado (Springer, Berlin, 1997)
26. H. Trotter, Proc. Am. Math. Soc. **10**, 545 (1959); M. Suzuki, Prog. Theor. Phys. **56**, 1454 (1976); R. Feynman and A. Hibbs *Quantum Mechanics and Path Integrals* (McGraw-Hill, 1965)
27. R. Bursill, T. Xiang and G. Gehring, J. Phys. Cond. Mat. **8**, L583 (1996); X. Q. Wang and T. Xiang, Phys. Rev. B **56**, 5061 (1997); N. Shibata, J. Phys. Soc. Jpn **66** 2221 (1997)
28. R. Bursill, T. Xiang and G. Gehring, J. Phys. C **8**, L583 (1996)
29. X. Wang and T. Xiang, Phys. Rev. B **56**, 5061 (1997)
30. N. Shibata and K. Ueda, J. Phys. Soc. Jpn. Vol.70, 3690 (2001)
31. N. Shibata, J. Phys. Soc. Jpn. **66**, 2221 (1997)
32. N. Shibata, J. Phys. A, **36** 381 (2003)
33. S. White and R. Martin, J. Chem. Phys. **110**, 4127 (1999); see also S. White in 2, Chap. 2.1.
34. S. White in Ref. 2, Chap. 2.1.
35. Ö. Legeza, J. Röder and B.A. Hess, Phys. Rev. B, **67**, 125114 (2003)
36. S. Daul, I. Ciofini, C. Daul and S. R. White, Int. J. Quantum Chem. **79**, 331 (2000)
37. M. A. Cazalilla and J. B. Marston, Phys. Rev. Lett. **88**, 256403 (2002)
38. P. Schmitteckert, Phys. Rev. B **70**, 121302 (2004)
39. J. Z. Lou, T. Xiang and X. Wang, Phys. Rev. Lett. **91**, 049701 (2003)
40. G. Vidal, Phys. Rev. Lett. **91**, 147902 (2003)
41. S. White and A. Feiguin, Phys. Rev. Lett. **93**, 076401 (2004)
42. K. Hallberg, Phys. Rev. B **52**, 9827 (1995)

43. T. Kühner and S. White, Phys. Rev. B **60**, 335(1999)
44. Y. Anusooya, S. Pati and S. Ramasesha, J. Chem. Phys. **106**, 1 (1997); S. Ramasesha, K. Tandon, Y. Anusooya and S. Pati, Proc. of SPIE, **3145**, 282 (1997); S. Ramasesha, Z. Shuai and J. Brédas, Chem. Phys. Lett. **245**, 224 (1995)
45. E. Jeckelmann, Phys. Rev. B **66**, 045114 (2002)
46. E. R. Gagliano and C. A. Balseiro, Phys. Rev. Lett. **59**, 2999 (1987).
47. G. Grosso and G. Partori Parravicini, in *Memory Function Approaches to Stochastic Problems in Condensed Matter*, Adv. in Chemical Physics, **62**, 133 (Wiley, N. Y., 1985)
48. Y. Nishiyama, Eur. J. Phys. B **12**, 547
49. C. L. Zhang, E. Jeckelmann and S. White, Phys. Rev. B **60**, 14092 (1999)
50. W. Q. Yu and S. Haas, Phys. Rev. B, **63**, 024423 (2000)
51. K. Okunishi *et al.*, Phys. Rev. B **64**, 104432 (2001)
52. T. Nunner, P. Brune, T. Kopp, M. Windt and M. Grüninger, cond-mat/0203472
53. K. Penc, K. Hallberg, F. Mila and H. Shiba, Phys. Rev. Lett. **77**, 1390 (1996); Phys. Rev. B, **55**, 15475 (1997)
54. S. S. Kancharla and C. J. Bolech, Phys. Rev. B **64**, 085119 (2001)
55. F. Essler, F. Gebhard and E. Jeckelmann, Phys. Rev. B **64**, 125119 (2001); Phys. Rev. Lett. **85**, 3910 (2000)
56. E. Jeckelmann, Phys. Rev. Lett. **91**, 089702 (2003)
57. Y.-J Kim *et al.*, Phys. Rev. Lett. **92**, 137402 (2004)
58. S. K. Pati *et al.*, Phys. Rev. B **60**, 7695 (1999)
59. T. D. Kühner, S. White and H. Monien, Phys. Rev. B **61**, 12474 (2000)
60. C. Raas, G. Uhrig and F. B. Anders, Phys. rev. B **69**, 041102(R) (2004)
61. G. Kotliar and D. Vollhardt, *Physics Today*, 53, March 2004
62. J.E. Hirsch and R.M. Fye, *Phys. Rev. Lett.* **56**, 2521 (1986)
63. M.J. Rozenberg, *Phys. Rev. B* **55**, R4855 (1997)
64. K. G. Wilson, *Rev. Mod. Phys.* **47**, 773 (1975)
65. R. Bulla, *Phys. Rev. Lett.* **83**, 136 (1999)
66. R. Bulla, T. Costi and D. Vollhardt, *Phys. Rev. B* **64**, 045103 (2001)
67. D. García, K. Hallberg and M. Rozenberg, to appear in Phys. Rev. Lett. (2004)
68. A. Georges *et al.*, *Rev. Mod. Phys.* **68**, 13 (1996)
69. Q. Si *et al.*, *Phys. Rev. Lett.* **72**, 2761 (1994)
70. M. J. Rozenberg, G. Moeller and G. Kotliar, *Mod. Phys. Lett. B* **8**, 535 (1994)
71. See F. Gebhard *et al.*, *Eur. Phys. J B* **36**, 491 (2003), for a previous attempt using a particular algorithm of the DMRG restricted to the small U metallic regime.
72. S. Nishimoto, F. Gebhard and E. Jeckelmann, J. Phys.: Cond. Matt. **16**, 7063, (2004)
73. A. Georges and G. Kotliar, *Phys. Rev. B* **45**, 6479 (1992)
74. X. Y. Zhang, M. J. Rozenberg and G. Kotliar, *Phys. Rev. Lett.* **70**, 1666 (1993), *Phys. Rev. B* **48**, 7167 (1993), A. Georges and W. Krauth Phys. Rev. B 48, 7167-7182 (1993).
75. G. Kotliar and D. Vollhardt, *Physics Today* **57**, 53 (2004)

FUNCTIONAL RENORMALIZATION GROUP IN THE 2D HUBBARD MODEL

CARSTEN HONERKAMP
Max Planck Institute for Solid State Research,
D-70569 Stuttgart, Germany
C.Honerkamp@fkf.mpg.de

Received 1 December 2004

We review recent developments in functional renormalization group (RG) methods for interacting fermions. These approaches aim at obtaining an unbiased picture of competing Fermi liquid instabilities in the low-dimensional models like the two-dimensional Hubbard model. We discuss how these instabilities can be approached from various sides and how the fermionic RG flow can be continued into phases with broken symmetry.

Keywords: Many-fermion systems; renormalization group; Hubbard model.

1. Introduction

Competing Fermi liquid instabilities are a general feature of low-dimensional interacting lattice electron systems. These instabilities are driven by long-lived particle-particle or particle-hole excitations near the Fermi surface (FS). They can lead to long range correlations of various types and a destruction of Fermi liquid behavior. In two and higher dimensions perturbative renormalization group methods are probably the least biased approach to study such systems at weak coupling and low energy scales in a transparent way.

In the last decade various functional renormalization group (fRG) methods [1,2,3,4,5] for many-fermion systems have been developed. These schemes are capable of capturing the interplay of various low-energy scattering processes within certain approximations. However, the practical experience shows that each scheme has advantages and disadvantages. This has led to the development of alternative flow schemes.[6,7] They provide the possibility to tackle the competing instabilities from different angles. Fortunately, it turns out that besides well-understood conceptual differences the flow of these schemes agrees quite favorably over a wider parameter range of models like the two-dimensional Hubbard model. With this understanding we hope that further improvements of the approach are on a firm basis.

This short review is organized as follows. First we outline the general formalism of the renormalization group for the one-particle irreducible vertex functions.[8,5,9] We explain how cutoff flow, temperature flow (T-flow) and interaction flow (IF) are obtained. Then we review some results for the two-dimensional Hubbard model.

Finally we describe a novel approach how the fermionic RG flow can be continued into phases with broken symmetry.

2. Functional RG for 1PI Vertex Functions

2.1. *General setup*

The class of models we study here is described by an action of the form

$$\begin{aligned} S(\psi,\bar\psi) &= \textstyle\sum_{k,s} \bar\psi_{k,s} Q(k)\, \psi_{k,s} \\ &\quad + \tfrac{1}{2N} T^3 \textstyle\sum_{\substack{k,k',q \\ s,s'}} V(k,k',k+q) \bar\psi_{k+q,s} \bar\psi_{k'-q,s'} \psi_{k',s'} \psi_{k,s} \,. \end{aligned} \tag{1}$$

Here $\bar\psi(k,s)$ and $\psi(k,s)$ are Grassmann fields representing fermions with wavevector $\vec{k}$, Matsubara frequency k_0 (we write $k=(k_0,\vec{k})$) and spin projection $s=\pm1/2$. $Q(k)$ denotes the quadratic part of the action, here given by $Q(k) = \left[-ik_0 + \epsilon(\vec{k})\right]$. For fermions on a 2D square lattice with nearest neighbor hopping t, next-nearest neighbor hopping t' and chemical potential μ, the dispersion is

$$\epsilon(\vec{k}) = -2t\,(\cos k_x + \cos k_y) - 4t' \cos k_x \cos k_y - \mu \,. \tag{2}$$

$V(k,k',q)$ defines the spin-rotationally invariant, frequency- and wavevector-conserving interaction between two fermions. For the repulsive onsite interaction of the Hubbard model $V(k,k',q) = U > 0$.

The connected correlation functions of the theory defined by (1) can be obtained by taking derivatives of the generating functional,

$$e^{-W(\xi,\bar\xi)} = \int \mathcal{D}\psi(k,s)\mathcal{D}\bar\psi(k,s)\, e^{-S(\psi,\bar\psi)+\sum_{k,s}\left[\bar\psi(k,s)\xi(k,s)+\bar\xi(k,s)\psi(k,s)\right]} \,. \tag{3}$$

For our purposes it is simpler to work with vertex functions, which are generated by the Legendre transform of $W(\xi,\bar\xi)$ with respect to $\phi = \delta W/\delta\xi$,

$$\Gamma(\phi,\bar\phi) = W(\xi,\bar\xi) - \sum_{k,s} \left[\bar\phi(k,s)\xi(k,s) + \bar\xi(k,s)\phi(k,s)\right] \,. \tag{4}$$

This functional can be expanded in monomials of its sources,

$$\Gamma(\phi,\bar\phi) = \sum_{m\geq 0} \frac{1}{m!} \sum_{\underline{K}} \gamma^{(m)}(\underline{K}) \Phi^m(\underline{K}) \,. \tag{5}$$

$\underline{K}$ is a multi-index which contains m frequencies, wavevectors, spins and Nambu indices of the Φ-fields. Nambu index + stands for $\bar\phi$ and $-$ for ϕ. The $\gamma^{(m)}$ are the m-point vertex functions of the theory. They are antisymmetric with respect to interchange of two particle coordinates.

The two-point vertex $\gamma^{(2)}$ is the inverse of the full propagator. For spin-rotational and U(1) invariance the antisymmetric four-point vertex for incoming particles k_1, s_1, k_2, s_2 and outgoing particles k_3, s_3, k_4, s_4 (the quantum numbers of particle

4 are dictated by the conservation laws) can be expressed in terms of a coupling function $V(k_1, k_2, k_3)$ by[5]

$$\gamma^{(4)}_{s_1 s_2 s_3 s_4}(k_1, k_2, k_3, k_4) = V(k_1, k_2, k_3)\delta_{s_1 s_3}\delta_{s_2 s_4} - V(k_2, k_1, k_3)\delta_{s_1 s_4}\delta_{s_2 s_3} \tag{6}$$

Now let us assume that the quadratic part of the action depends continuously on a parameter ℓ while the interaction terms does not contain ℓ. Then we can derive an exact equation[8,5,9] for the change of $\Gamma_\ell(\phi, \bar{\phi})$ when ℓ is varied by taking the ℓ-derivative of Eq. 4. Inserting the field expansion (5) yields an infinite hierarchy of differential equations. This set of equations defines the renormalization group flow of the 1PI vertex function with respect to the parameter ℓ. In order to make this system tractable, one has to truncate it at some point. Here all 1PI vertex functions with $m > 4$ are set to zero. This is a first approximation that is common to all RG schemes discussed below. The remaining equations for $\gamma^{(2)}_\ell$ and $\gamma^{(4)}_e ll$ are shown graphically in Fig. 1.

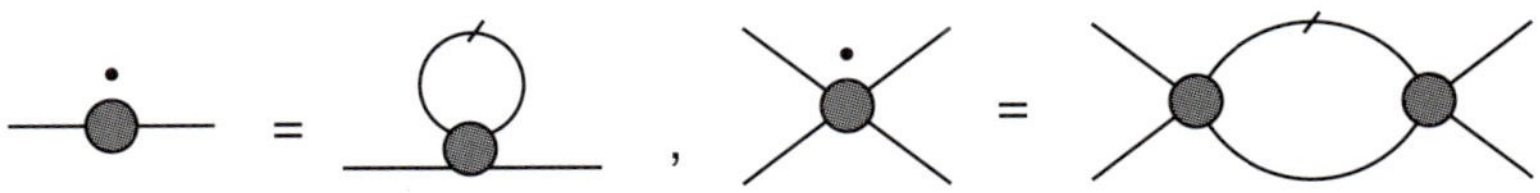

Fig. 1. RG equations for the two-point and the four-point vertex. The slashed line denotes a single-scale propagator $S_\ell(p)$. The one-loop graph for the four-point vertex $\gamma^{(4)}_\ell$ includes particle-particle and particle-hole contributions. With our truncation $\gamma^{(m)}_\ell = 0$ for $m \geq 6$ the feedback of the $\gamma^{(6)}_\ell$ on $\gamma^{(4)}_\ell$ is neglected.

We emphasize that in the exact hierarchy the right hand side contains only one-loop diagrams. This is one of the blessings of the approach which in certain cases even allows for a non-perturbative treatment.[13] Higher-loop diagrams are included via the ℓ-dependence of the vertices. They can be made explicit by reinserting the integrated flow of the vertices into the right hand side.[10] The equation for the coupling function $V_\ell(p_1, p_2, p_3)$ with $p_4 = p_1 + p_2 - p_3$ is[5]

$$\frac{d}{d\ell} V_\ell(p_1, p_2, p_3) = \mathcal{T}_{PP,\ell} + \mathcal{T}^{d}_{PH,\ell} + \mathcal{T}^{cr}_{PH,\ell} \tag{7}$$

with the one-loop particle-particle contributions $\mathcal{T}_{PP,\ell}$ and the two different particle-hole channels $\mathcal{T}^{d}_{PH,\ell}$ and $\mathcal{T}^{cr}_{PH,\ell}$ where

$$\mathcal{T}_{PP,\ell}(p_1, p_2; p_3, p_4) = \\ - \int dp\, V_\ell(p_1, p_2, p)\, L(p, -p + p_1 + p_2)\, V_\ell(p, -p + p_1 + p_2, p_3) \tag{8}$$

$$\mathcal{T}^{d}_{PH,\ell}(p_1, p_2; p_3, p_4) = \\ - \int dp \Big[-2V_\ell(p_1, p, p_3)\, L(p, p + p_1 - p_3)\, V_\ell(p + p_1 - p_3, p_2, p) \\ + V_\ell(p_1, p, p + p_1 - p_3)\, L(p, p + p_1 - p_3)\, V_\ell(p + p_1 - p_3, p_2, p)$$

$$+V_\ell(p_1,p,p_3)\,L(p,p+p_1-p_3)\,V_\ell(p_2,p+p_1-p_3,p)\Big] \tag{9}$$

$$\mathcal{T}^{cr}_{PH,\ell}(p_1,p_2;p_3,p_4) =$$
$$-\int dp\,V_\ell(p_1,p+p_2-p_3,p)\,L(p,p+p_2-p_3)\,V_\ell(p,p_2,p_3) \tag{10}$$

In these equations, the product of the two internal lines in the one-loop diagrams is

$$L(p,p') = S_\ell(p)W^{(2)}_\ell(p') + W^{(2)}_\ell(p)S_\ell(p') \tag{11}$$

with the so-called single-scale propagator

$$S_\ell(p) = -W^{(2)}_\ell(p)\left[\frac{d}{d\ell}Q_\ell(p)\right]W^{(2)}_\ell(p)\,. \tag{12}$$

Here, $W^{(2)}_\ell(p)$ denotes the full Green's function at RG scale ℓ.

2.2. *Momentum-shell RG, temperature flow and interaction flow*

So far the setup of the fRG equations has been very general. Now we specify the flow parameter ℓ. We discuss three different choices (although many more are possible). The guiding principle for a good flow parameter is that it should enable us *a)* to approach a specific singularity in the perturbation expansion in a controlled way and *b)* to include all other (possibly also singular) tendencies during the flow. Then the RG will give a more realistic picture than perturbation expansions which single out one dominant channel. In bubble or ladder summations, the singularities normally arise due to a pile-up of logarithms. These are roughly given by $g\log[W/\max(T,\Lambda)]$ with a coupling constant g, bandwidth W and lower energy cutoff Λ. Hence, if we want to build in such dangerous terms step by step, we can either vary Λ, the temperature T, or the coupling g. This is made more precise below.

Cutoff-RG

Here we introduce a cutoff-function $C_\Lambda[\epsilon(\vec{k})]$ into the quadratic part of the action,

$$Q_\Lambda(k) = T\sum_{ik_0,\vec{k}} C_\Lambda[\epsilon(\vec{k})]\,\bar{\psi}_k\left[-ik_0+\xi(\vec{k})\right]\psi_k\,. \tag{13}$$

$C_\Lambda[\epsilon(\vec{k})]$ is very large for $|\epsilon(\vec{k})| \le \Lambda$ and $C_\Lambda[\epsilon(\vec{k})] = 1$ for $|\epsilon(\vec{k})| > \Lambda$ such that modes below Λ are not integrated over in the functional integral. In practice one mainly needs the inverse $C^{-1}_\Lambda[\epsilon(\vec{k}]$, which can be chosen conveniently as a sharp step function for analytical manipulations or as a smoothed step function for numerical treatments. The full Green's function is suppressed for modes with $|\epsilon(\vec{k})| \le \Lambda$. A frequency cutoff (or combinations of frequency and band energy) falls into the same class of cut-off RG. For our $\vec{k}$-space resolved approach described below a pure energy-cutoff is however most convenient. The momentum-shell RG is the widely used standard[1] for many-fermion systems. It can also be derived from other exact RG equations.[11,3,12,2,4] It approaches Fermi liquid instabilities for fixed model parameters and fixed temperature. One of the successes of the method in the weakly coupled 2D Hubbard model are the clear signatures for d-wave superconductivity over a wide parameter range.[3,4,16]

A serious drawback of the cutoff RG is however the non-uniformity in the RG scale at which one-loop particle-hole (PH) processes with different wavevector transfers $\vec{q}$ are included at low T. Typically PH pairs with a particle $\vec{k}$ and a hole $\vec{k}+\vec{q}$ and large $|\vec{q}| \sim k_F$ are integrated out at all scales Λ in the flow, depending on how close $\vec{k}$ and $\vec{k}+\vec{q}$ are to the Fermi surface. In contrast with this, PH processes with small $\vec{q}$ can only occur in the vicinity of the Fermi surface. For $T \to 0$ the support of the PH bubble for $\omega = 0, \vec{q} \to 0$ shrinks to a temperature-smeared δ-function on the FS with width $\sim T$. Thus, the $\vec{q} \to 0$ PH-modes are integrated out only for cutoff $\Lambda \leq T$, even if the density of states is divergent, and these processes give a singular contribution for $T \to 0$. However, in a coupled flow with a lot of other tendencies, e.g. the omni-presence of Kohn-Luttinger superconducting instabilities, the flow normally diverges *before* we get down to $\Lambda \sim T$. The flow has to be stopped and the $\vec{q} \to 0$ PH pairs did not have the chance to contribute to the flow *by construction*. Therefore, the cutoff-flow is still biased. Other approaches are needed to study the influence of the $\vec{q} \to 0$ particle-hole excitations.

The following two schemes work without a cutoff and allow for a uniform inclusion of the various one-loop processes.

Temperature flow

The temperature-flow scheme[6] uses the temperature T appearing in front of the frequency sum and in the Matsubara frequencies as flow parameter. Before the T-derivative can be taken, the fermionic fields need to be rescaled in order to remove the T^3 term appearing in front of the interaction term in the original action. Then the strategy is as follows. The theory is defined at some high T $\sim T_0$ in terms of its two-point and four-point vertex. Here, we simply choose the free propagator and the local Hubbard repulsion as the initial values at T_0. This should be reasonable for sufficiently high $T_0 \sim$ bandwidth, as all perturbative corrections decay with a negative power of T. In the approximate version used below without $\gamma_T^{(6)}$ and self-energy corrections, the right hand side of the flow equation for the interaction is just given by the T-derivatives of the one-loop diagrams.

Interaction flow

In the interaction flow scheme[7] we first multiply Q with a scale factor $1/g$ and split it in two, yielding

$$Q_g = T \sum_{i\omega_n, \vec{k}} \bar{\psi}_k g^{-1/2} \left[-i\omega_n + \xi(\vec{k}) \right] \psi_k g^{-1/2} \, . \tag{14}$$

g will be the flow parameter. We can absorb the factor $1/g$ in rescaled fields $\tilde{\psi}, \bar{\tilde{\psi}}$ defined as $\tilde{\psi} = g^{-1/2}\psi$. With this the interaction term gets an extra factor g^2 when written in terms of the new fields:

$$V_g^{(4)} = \frac{1}{2N} \sum_{\substack{k,k',k+q \\ s,s'}} g^2 V(k,k',k+q) \bar{\tilde{\psi}}_{k+q,s} \bar{\tilde{\psi}}_{k'-q,s'} \tilde{\psi}_{k',s'} \tilde{\psi}_{k,s} \,. \tag{15}$$

We observe that changing the scale factor $1/g$ in Q_g corresponds to changing the strength of the bare interactions. The rescaled fermions $\tilde{\psi}$, $\bar{\tilde{\psi}}$ describe a system with a bare interaction strength $g^2 V$. Now we can start at $g = 0^+$, i.e. at infinitesimally small bare interaction, and use the flow equations to integrate up to the desired bare interaction, reached at $g = 1$. We can also stop the flow at any other value of g, with the functions $g\Sigma$ and $g^2 V_g(k,k',k+q)$ being the self-energy and interacting vertex function for the bare interaction $g^2 V(k,k',k+q)$. We call this the *interaction flow* (IF). Related schemes have been proposed by Polonyi[14] and Meden.[15]

In the IF, singularities on the right hand side of the flow equation are not regularized by the flow parameter. Thus, the IF scheme has to be performed at $T > 0$, when the individual one-loop diagrams are bounded. The strength of the bare interaction is increased continuously, and in the course of the interaction flow potential singularities are approached from below.

Fig. 2 a) illustrates how the various methods detect perturbative singularities from different directions in parameter space. We emphasize that the T-flow and the IF scheme do not correspond to viewing a system on different length scales.

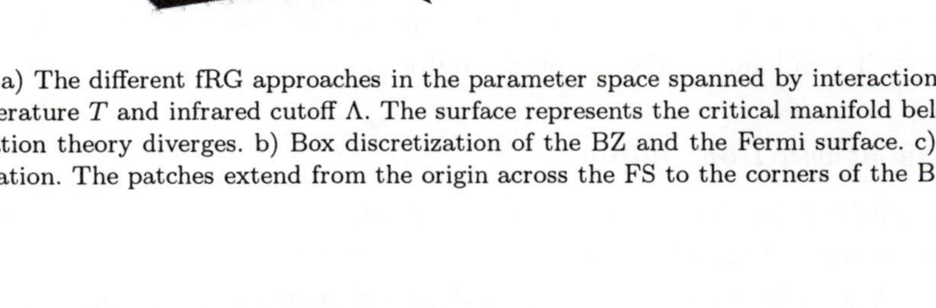

Fig. 2. a) The different fRG approaches in the parameter space spanned by interaction strength U, temperature T and infrared cutoff Λ. The surface represents the critical manifold below which perturbation theory diverges. b) Box discretization of the BZ and the Fermi surface. c) N-patch discretization. The patches extend from the origin across the FS to the corners of the BZ.

3. Application to the 2D Hubbard Model

Here we review results[10,16] obtained for the 2D t-t' Hubbard model with next-nearest neighbor hopping $t' \neq 0$. The case $t' = 0$ was studied with related cutoff-RG schemes e.g. by Zanchi and Schulz[3] and Halboth and Metzner.[4]

3.1. *Calculational issues and approximations*

All fRG approaches to the Hubbard model have so far neglected the frequency dependence of the interaction vertex by calculating the static part and approximating all other vertices with the static one. This can be argued to be good as the initial interactions are frequency independent, and the corrections to the leading divergences in the static components are small by power counting. The next approximation is the neglect of the self-energy corrections. The consistency of this can be checked by calculating the self-energy from the flow of the couplings. With these approximations and the initial truncation of the hierarchy one is left with the flow of the coupling function $V_\ell(\vec{k}_1, \vec{k}_2, \vec{k}_3)$. This function of three wavevectors can either be discretized directly on a grid in the Brillouin zone (see Fig. 2 b)), or, more physically motivated, in the so-called N-patch scheme, first used in this context by Zanchi and Schulz.[3] This scheme takes advantage of the fact that for the standard Fermi liquid instabilities the wavevector dependence of the coupling function $V(\vec{k}_1, \vec{k}_2, \vec{k}_3)$ has a relevant part, which is the dependence of the processes between different points $\vec{k}_1^F, \vec{k}_2^F, \vec{k}_3^F$ on the FS when the $\vec{k}_i^F$ are varied around it. Furthermore there is an irrelevant part, which is the dependence in the directions orthogonal to the FS. Hence, one calculates the $V(\vec{k}_1, \vec{k}_2, \vec{k}_3)$ for $\vec{k}_1, \vec{k}_2, \vec{k}_3$ on the FS and treats it as piecewise constant when $\vec{k}_1, \vec{k}_2$, and $\vec{k}_3$ move within elongated patches stretching from the origin of the BZ to the $(\pm\pi, \pm\pi)$-points (see Fig. 2 c)).

Then, the RG flow is started at an initial scale Λ_0 for the cutoff RG, at temperature T_0 for the T-flow, or initial interaction strength $g^2 = 0^+$ for the interaction flow, respectively. The initial interaction is chosen as $V_\ell(\vec{k}_1, \vec{k}_2, \vec{k}_3) = U$ $(= g^2 U$ for the IF$)$. What is typically encountered at low T is a *flow to strong coupling*, where for a certain flow parameter ℓ_c one or several components of $V_\ell(\vec{k}_1, \vec{k}_2, \vec{k}_3)$ become large. At that point the approximations break down, and the flow has to be stopped. Physical information is obtained from susceptibilities that can be calculated from the flow and directly from the $\vec{k}$-dependence of $V_\ell(\vec{k}_1, \vec{k}_2, \vec{k}_3)$. For example, if only processes with $\vec{k}_1 + \vec{k}_2 = 0$ become large in magnitude (typically strongly attractive), the flow to strong coupling corresponds to a Cooper instability, indicating a dominant tendency to form a superconducting state in the parameter regime beyond the divergence. The symmetry of the superconducting pairing can be read off from the $\vec{k}_1$-dependence of the coupling function.

3.2. *Flow at van Hove filling*

First we consider the flow at the so-called van Hove (VH) filling, where the FS is kept fixed at the VH singularities of the free dispersion at $\vec{k} = (\pi, 0)$ and $(0, \pi)$. We change the band filling by varying the next-nearest neighbor hopping t'.

Using the T-flow scheme one finds three regimes.[6] In the first regime for small $|t'|$ the FS is properly nested and there is a antiferromagnetic (AF) spin density wave regime. Then, for increasing $|t'|$, the $d_{x^2-y^2}$-wave superconducting tendencies get stronger and there is a regime where the corresponding susceptibility diverges most strongly. Around $t' = -0.33t$, the flow to strong coupling is suppressed by orders of magnitude. On the other side of this potential quantum critical point the flow is dominated by ferromagnetic ordering tendencies. The same qualitative picture is found with the IF scheme,[7] although in this approach the d-wave tendencies seem to be weaker by a certain degree, most probably due to the unfavorable box-discretization (Fig. 2 b)) used. Also, a parquet approach focusing on the VH regions gives the same order of instabilities.[19] On the contrary, the cutoff RG misses the ferromagnetic regime completely. The reason is the above-mentioned inability of the cutoff RG to treat small-$\vec{q}$ particle-hole pairs on equal footing with other tendencies. Instead it produces a low-T_c Kohn-Luttinger instability in the singlet channel.

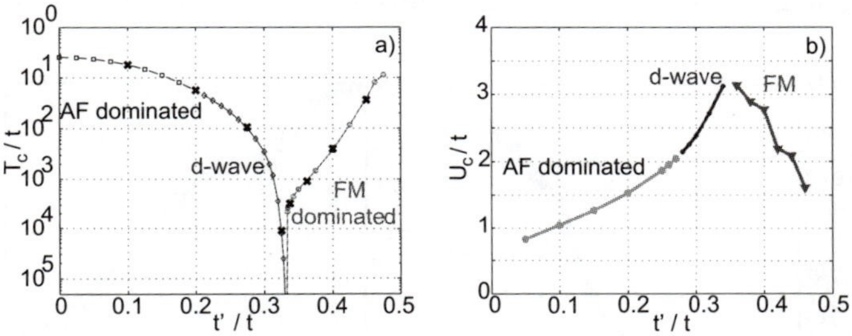

Fig. 3. a) Critical temperatures T_c for the flow to strong coupling and dominant correlations at the van Hove filling, obtained with T-flow RG for $N = 48$ patches and $U = 3t$. b) Critical bare interaction strengths for the flow to strong coupling, obtained with IF scheme at $T = 0.001t$ and the van Hove filling and 12×12 box discretization.

3.3. *Flow with fixed t': Saddle point regime*

Next we briefly sketch the tentative phase diagrams from the RG flow for fixed $t' = -0.3t$ as obtained with the cutoff RG.[16,10] These values of t' might be relevant for the high-T_c problem. However, the Mott insulating state of the high-T_c compounds can not be obtained with our weak coupling approach. Hence, our results cannot be directly compared to the cuprates. We can however read off some tendencies. For these parameter values there are still quantitative differences between the various RG schemes. These mainly concern the location of the boundaries and the numbers for critical scales. The order of the regimes however is the same for all approaches.

First let us consider band fillings $n > 1$ which correspond to electron-doping in the high-T_c language. Here the RG finds just two phases when n is varied. One is an AF nesting regime with relatively large critical scales closer to half filling. Increasing the particle number reduces nesting, and finally (π, π) particle-hole become impossible at low T (see FS in the left plot of Fig. 4). Then (or at somewhat lower doping, depending on the RG scheme) the AF tendencies become cut off. What remains is a low-scale d-wave superconducting instability.

On the hole-doped side with $n < 1$ the FS is close to the VH points and the *saddle point regions* around $(\pi, 0)$ and $(0, \pi)$ play a dominant role. Between an AF nesting regime near half filling and a d-wave regime further away we find another regime, the so-called *saddle point regime*. Here several channels, most prominently the AF spin density wave, the d-wave Cooper, and also the d-density wave channel grow with comparable strength. All these channels reinforce each other mutually.[16] Hence, it is not obvious that the system goes into a long range ordered phase. A similar flow arises in Hubbard two-leg ladder systems. There the ground state is known to be a short-range correlated Mott insulator. Therefore it is conceivable that the saddle point regions of the 2D system will adopt a similar gapped strong coupling state with a truncated FS.[17,16] We have elaborated on this idea recently by exact diagonalization of effective Hamiltonians suggested by the RG flow.[18]

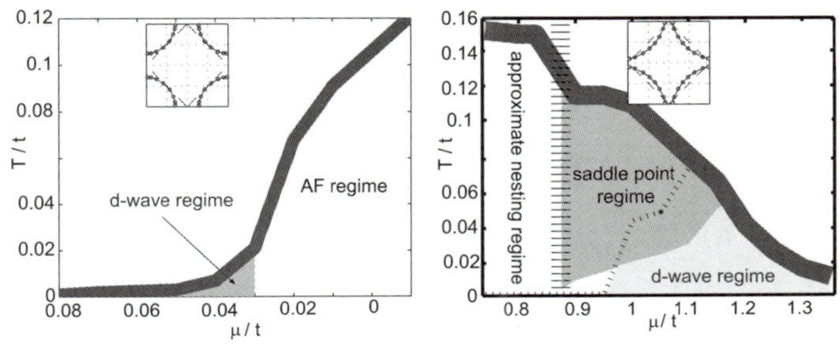

Fig. 4. Summary of the RG flow for fixed $t' = -0.3t$ obtained with N-patch cutoff RG. The left plot is for electron doping, i.e. band filling $n > 1$, while the right plot is for hole doping, $n < 1$. The band is half filled for $\mu \approx -0.65t$. The insets show the location of the Fermi surfaces for the different chemical potentials. Above the thick lines the flow to strong coupling is cut off by temperature.

4. Continuation of the Flow through the Instability

So far we have only discussed RG flows where the effect of self-energy corrections has been neglected. This approximation will certainly break down when the couplings get large and it is desirable to include at least parts of the self-energy.

There have been several fRG studies[10,20,21,22,23] which gave some insights how the normal state self-energy is altered by the flow of the interactions. Yet in these works the RG flow still had to be stopped when the coupling constants became large. Physically one may expect that in many cases the opening of a gap in the fermionic spectrum should regularize this apparent divergence. Recently we have analyzed this process in the reduced BCS model using the 1PI cutoff-RG.[25] In this model only particle pairs with zero total momentum interact with an attractive coupling constant. A theory with two- and four-point vertices only can be seen to be exact. The essential step is to include a small symmetry-breaking off-diagonal part Δ_0 into the initial condition for the self-energy. Now also anomalous propagators appear in the flow when the cutoff is lowered. They generate a anomalous interaction vertex which has 4 incoming or 4 outgoing lines. The flow equations for the self-energy and the interaction vertices as shown in Fig. 1 can be integrated down to cutoff $\Lambda = 0$ without a divergence of the interactions. A necessary point is a reorganization of the flow for the four-point vertex proposed recently by Katanin.[24] The $\Lambda = 0$-result for the anomalous self-energy approaches the BCS result for $\Delta_0 \to 0$. With initial $\Delta_0 > 0$ the flow of the four-point vertex remains finite. Only for $\Delta_0 \to 0$, a combination of normal and anomalous vertices diverges like $1/\Delta_0$. This corresponds to the vanishing mass of the Goldstone boson when the U(1)-symmetry is restored.

So far this scheme has only been applied to the reduced BCS model. It will however also work for more general cases where the interplay of particle-particle and particle-hole processes drives the system towards a superconducting instability. If a nonzero Δ_0 is included in the initial conditions, the flow down to $\Lambda = 0$ will converge to a state with a gapped spectrum and a controlled magnitude of the interactions. The strength of this scheme is that no mean-field decoupling of the interactions is required. In particular, the attractive interaction which goes into the equation for the anomalous self-energy is generated automatically during the flow to all orders in the bare interaction. This should allow for a much more precise investigation of the phase diagrams of weakly interacting fermion systems.

Acknowledgments

I thank Manfred Salmhofer for sharing many of his insights with me and T. Maurice Rice for motivation. Daniel Rohe is acknowledged for comments on the manuscript.

References

1. R. Shankar, Rev. Mod. Phys. **66**, 129 (1994).
2. M. Salmhofer, Renormalization (Springer, Berlin, 1998).
3. D. Zanchi, and H.J. Schulz, Europhys. Lett. **44**, 235 (1997); Phys. Rev. B **61**, 13609 (2000).
4. C. J. Halboth, and W. Metzner, Phys. Rev. B **61**, 7364 (2000); Phys. Rev. Lett. **85**, 5162 (2000).
5. M. Salmhofer, and C. Honerkamp, Prog. Theor. Physics **105**, 1 (2001).
6. C. Honerkamp, and M. Salmhofer, Phys. Rev. B **64** 184516 (2001); Phys. Rev. Lett. **87**, 187004 (2001).
7. C. Honerkamp, D. Rohe, S. Andergassen and T. Enss, cond-mat/0403633, to appear in PRB.
8. C. Wetterich, Phys. Lett. B**301**, 90 (1993).
9. P. Kopietz and T. Busche, Phys. Rev. B **64** , 155101 (2001); P. Kopietz, Nucl.Phys. B **595**, 493 (2001).
10. C. Honerkamp, Euro. Phys. J. B **21**, 81 (2001).
11. J. Polchinski, Nucl. Phys. B **231**, 269 (1984).
12. S. W. Tsai and J. B. Marston, Can. J. Phys. **79**, 1463 (2001).
13. J. Berges, N. Tetradis, C. Wetterich, Phys. Rep. **363**, 223 (2002).
14. J. Polonyi, Act. Phys. Slovacia. **52**, 295 (2002) (cond-mat/0205040).
15. V. Meden, unpublished.

16. C. Honerkamp, M. Salmhofer, N. Furukawa and T. M. Rice, Phys. Rev. B **63**, 035109 (2001).
17. N. Furukawa, M. Salmhofer, and T.M. Rice, Phys. Rev. Lett. **81**, 3195 (1998).
18. A. Läuchli, C. Honerkamp, and T.M. Rice, Phys. Rev. Lett. **92**, 037006 (2004)
19. V. Yu. Irkhin, A. A. Katanin, and M. I. Katsnelson, Phys. Rev. B**64**, 165107 (2001)
20. D. Zanchi, Euro. Phys. Lett. **55**, 376 (2001).
21. C. Honerkamp and M. Salmhofer, Phys. Rev. B **67**, 174504 (2003).
22. A. A. Katanin and A. P. Kampf, Phys. Rev. Lett. **93**, 106406 (2004). A. A. Katanin, A. P. Kampf, V. Yu. Irkhin, cond-mat/0407473.
23. D. Rohe and W. Metzner, cond-mat/0406164
24. A.A. Katanin, Phys. Rev. B **70**, 115109 (2004)
25. M. Salmhofer, C. Honerkamp, W. Metzner, and O. Lauscher, cond-mat/0409725.

QUANTUM PHASE TRANSITIONS AND EVENT HORIZONS: CONDENSED MATTER ANALOGIES

GEORGE CHAPLINE
Lawrence Livermore National Laboratory,
Livemore, CA 94550

Received 1 December 2004

Although it has been generally believed that classical general relativity is always correct for macroscopic length scales, certain predictions such as event horizons and closed time-like curves are inconsistent with ordinary quantum mechanics. It has recently been pointed out that the event horizon problem can be resolved if space-time undergoes a quantum phase transition as one approaches the surface where general relativity predicts that the redshift becomes infinite. Indeed a thought experiment involving a superfluuid with a critical point makes such a suggestion appear plausible. Furthermore the behavior of space-time near an event horizon may resemble quantum phase transitions that have been observed in the laboratory. For example, the phenomenology of metamagnetic quantum critical points in heavy fermion materials resembles the behavior expected, both in terms of time standing still and the behavior of quantum correlation functions. Martensitic transformations accompanied by non-adiabatic changes in the electronic wave function are also interesting in this connection.

Keywords: Event horizon; quantum phase transition.

1. Quantum Mechanics and Space-Time

Although the early experimental successes of general relativity (GR) naturally led to the assumption that the predictions of classical general relativity should always be taken seriously, it turns out that general relativity has a serious disease; namely, general relativity is not always consistent with quantum mechanics. The fundamental reason for this inconsistency is that the formalism of quantum mechanics requires a universal time for its definition, whereas the classical Einstein equations often lead to space-times, e.g. any space-time with rotation, with no universal time.

Essentially every aspect of quantum mechanics requires a universal time. For example, what time does one mean when one writes down Schrodinger's equation. More subtle arguments involve the existence of non-local correlations. These non-local correlations can in fact exist over cosmological distances, as in Wheeler's delayed choice experiment. In Wheeler's thought experiment collapse of the wave function is required to occur over vast distances when an observer on earth points his telescope to view gravitationally lensing of a distant galaxy.

The validity of ordinary quantum mechanics seems to favor at a minimum space-times where it is possible to introduce a synchronous coordinate system. In such a coordinate system the off-diagonal components of the metric g_{0i} have been transformed to zero, and there is a universal time coordinate. If one visualizes the vacuum as analogous to the ground state of a condensed matter system and ordinary matter as analogous to excited states of this system, then it follows that the atoms in the condensed matter system must move without collective rotation. Indeed, in a coordinate system that is comoving with the collective motion of the atoms the contravariant 4-velocity is just $u^0 = 1$ and $u^i = 0$. In a synchronous coordinate system the covariant 4-velocity would then be $u^0 = 1$ and $u^i = 0$. Consequently, in a non-comoving coordinate system the collective velocity $\mathbf{v}$ would satisfy $\nabla \times \mathbf{v} = 0$. This condition suggests that an appropriate model for the vacuum of spacetime is a superfluid because the collective excitations of such a system are naturally irrotational.

In the following we will focus on superfluids as models for the vacuum of spacetime. It is a fortiori true that there is a universal time for such vacua. Indeed, as was first shown by Bogoliubov the phenomenon of superfluidity is a consequence of non-local quantum correlations between atoms in a condensate ground state. As noted above a universal time is required to make logical sense of such non-local quantum correlations.

2. A Thought Experiment

A simple thought experiment makes it clear why it is wrong to assume that classical GR is always correct on macroscopic length scales. Imagine that a Bose superfluid is condensed in a tall vertical cylinder. Now suppose that as a result of the increasing pressure in the superfluid as a function of depth the speed of sound vanishes at a certain depth inside the cylinder. What is noteworthy about this setup is that the behavior of sound waves near to the critical surface where the speed of sound vanishes is both well defined, and up to a certain distance from the critical surface, qualitatively indistinguishable from the classical behavior of light outside the event horizon of a Schwarzschild black hole.

It follows from the Schrodinger equation for a superfluid that the dispersion relation for small amplitude waves approaching a critical surface in a superfluid will have the form

$$\hbar\omega = \sqrt{(\hbar v_s q)^2 + \left(\frac{\hbar^2 q^2}{2m}\right)^2} \tag{1}$$

where v_s is the velocity of sound in the superfluid. As developed previously in a simple superfluid model for spacetime,[1] the event horizon of a classical black hole is replaced by a surface where the velocity of sound vanishes. However, in contrast with the behavior of waves or particles as they cross the event horizon of a classical black hole, the sound waves in our thought experiment would not pass through the critical surface in an uneventful way. There are two effects in particular that will be important for us. First, the frequency of the waves will become a quadratic function of wave number as they approach the critical surface, and secondly above a certain frequency $Q0$ that depends on the strength of the 4-wave interactions,[2] the waves will become unstable as they cross the critical surface.

Within the framework of classical general relativity nothing remarkable is expected to happen to an observer falling through the event horizon of a large black hole other than the curious circumstance that the observer can no longer communicate with the outside world. Although this prediction has been widely accepted in the physics community, it is inconsistent with quantum mechanics because it conflicts with the need for a universal time to define Schrodinger's equation.

3. Quantum Critical Behavior in Solids as a Model for an Event Horizon

The conceptual model for an event horizon just discussed invokes a Bose superuid with a critical point as a function of density where the speed of sound vanishes. Unfortunately, with the possible exception of helium atoms on a surface, there is no laboratory example of such a superfluid. On the hand, the phenomenon of time "standing still" at a quantum critical point (QCP) has been observed in the laboratory in a variety of solid materials. This effect was firrst seen in neutron scattering from CeCu doped with Au near a gold concentration of 10 percent, where the system exhibits non-Fermi liquid behavior.[3] Actually a drastic slowing down of quantum fluctuations may be a general characteristic of quantum critical systems. In addition, the behavior of the specific heat of solids near to QCPs is generally consistent with what might be expected for vacuum fluctuations in a quantum field theory near to an event horizon. As an alternative to a real superfluid model for an event horizon one might consider solid state systems with superfluid-like phases.

For example, quantum XY magnets or the low temperature behavior of atoms in an optical lattice. Both these systems have critical points where the speed of linearly dispersing collective excitations goes to zero. In the case of XY magnets the critical point is the isotropic Heisenberg quantum magnet, where the dispersion relation for spin waves changes from being linear to quadratic. Unfortunately, in neither of these systems does the superfluid-like phase continue beyond the critical point as envisioned in our thought experiment. In the case of the XY magnet the XY phase is replaced by an Ising phase where the spins are aligned along the z-axis. In the case of atoms in an optical lattice the superfluid phase undergoes a transition to a "Mott insulator" phase. In neither case does the phase on the other side of the critical point match our expectations for what the interior space-time should be like inside an event horizon. It is of course conceivable that the event horizon really is a first order phase transition. However, my prejudice for the quantum phase transition of space-time near an event horizon is that, completely analogous to a liquid-gas critical point, the nature of the quantum order is similar on both sides of the event horizon. A nice consequence of having the same kind of quantum order on both sides of the transition is that classical GR can be used to describe the interior of compact astrophysical objects, and the only difference between the interior and exterior is that interior has a much higher vacuum energy. One example of a class of solid materials with QCPs with characteristics similar to our hypothetical superfluid are heavy fermion meta-magnets, whose H vs M curves have the characteristic Van der Waals form. Examples of such materials include Sr3RuO7 and Ca2RuO4. Unfortunately these materials are not good models for space-time because they are metallic and so the analog of light waves, i.e. spin wave excitations, are strongly damped.

A perhaps more interesting class of QCPs is provided by materials with "continuous" martensitic transitions that can be driven to zero temperature with the application of pressure, and where the martensitic deformation is accompanied by a 2nd order electronic phase transition. Examples include AuZn alloys, the pyrochlore Cd2Re2O7,[4] and various ferroelectric materials. The interesting feature here is that in a certain direction the speed of sound vanishes near the transition. The hole doped high Tc cuprates may also be in this "universality class".

4. Acknowledgments

This work was performed (in part) under the auspices of the U.S. Department of Energy by University of California Lawrence Livermore National Laboratory under contract No. W-7405-Eng-48.

References

1. G. Chapline, E. Hohlfeld, R. Laughlin, and D. Santiago, Phil. Mag. B281, 235 (2001).
2. G. Chapline, Int. J. of Mod. Phys. A18, 3587 (2003).
3. A. Schroder, Nature 407, 351 (2000).
4. I. A. Sergienko et al., "Metallic Ferroelectricity in the Pyrochlore Cd2Re2O7", cond-mat/0402171.

SPIN–CHARGE SEPARATION AND TOPOLOGICAL PHASE TRANSITIONS IN AHARONOV–BOHM RINGS OF INTERACTING ELECTRONS

B. NORMAND

Département de Physique, Université de Fribourg, CH-1700 Fribourg, Switzerland
bruce.normand@unifr.ch

K. HALLBERG

Centro Atómico Bariloche and Instituto Balseiro, Comisión Nacional de Energía Atómica, 8400 Bariloche, Argentina

A. A. ALIGIA

Centro Atómico Bariloche and Instituto Balseiro, Comisión Nacional de Energía Atómica, 8400 Bariloche, Argentina

A. P. KAMPF

Institut für Physik, Theoretische Physik III, Elektronische Korrelationen und Magnetismus, Universität Augsburg, 86135 Augsburg, Germany

Received 12 November 2004

We investigate the properties of strongly correlated electronic models on a flux-threaded ring connected to semi-infinite free-electron leads. The interference pattern of such an Aharonov-Bohm ring shows sharp dips at certain flux values, determined by the filling, which are a consequence of spin-charge separation in a nanoscopic system. The conductance through such a molecule or nanodevice is related directly to its spectroscopic properties, opening new experimental possibilities for probing the phenomenology of strongly interacting systems. As a further example, for a ring described by the half-filled ionic Hubbard model we show that the weight of the first conductance peak as a function of gate voltage or external flux allows one to identify the topological charge transition between a correlated insulator and a band insulator.

Keywords: Aharonov-Bohm ring; spin–charge separation; topological phase transition.

1. Introduction: Aharonov–Bohm Ring Geometry

Modern fabrication technology is essentially capable of preparing nanostructures to order, and of contacting and measuring systems of sizes down to the molecular scale. Recent investigations of nanoscopic structures have included quantum dot arrays, quantum dot chains, quantum dot molecules, mesoscopic rings, single molecules, nanocrystals and nanotubes. Here we consider the detection of certain phenomena which arise purely due to many–body correlation effects in such systems.

Strong electronic correlations invalidate the conventional quasiparticle description of Fermi liquids, and in one dimension (1D) lead to a "fractionalisation" of electronic excitations into separate spin and charge modes.[1] Spin–charge separation (SCS) has been sought experimentally in a variety of materials[2] and studied numerically in small systems.[3,4] The concept of the topological phase transition, in this case from a band insulator to a correlated insulator, is illustrated by a class of 1D systems described by the ionic Hubbard model (IHM),[5] and is relevant in a number of organometallic compounds.

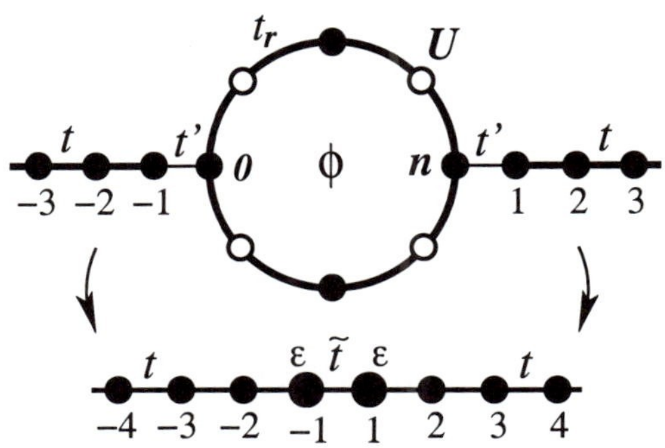

Fig. 1. Schematic representation of an interacting system on a ring connected by links t' to free-electron leads and threaded by a magnetic flux ϕ.

We study these phenomena by considering the transmittance of a flux-threaded, or Aharonov–Bohm (AB), ring (Fig. 1) of strongly interacting electrons attached to free-electron leads. This geometry was used to investigate the consequences of SCS in a Luttinger liquid.[4] The system can be considered as a chain containing two impurities (sites -1 and 1 in Fig. 1), and for small t' the transmittance is given by

$$T(\omega, V_g) = \frac{4\Gamma^2(\omega)|t_{\text{eff}}|^2}{|(\frac{\omega}{m} - \Delta\epsilon_{-1} + i\Gamma(\omega))(\frac{\omega}{m} - \Delta\epsilon_1 + i\Gamma(\omega)) - |t_{\text{eff}}|^2|^2}, \tag{1}$$

where the quantities

$$\Delta\epsilon_{-1}(\omega) = t'^2 g_{00}(\omega), \quad \Delta\epsilon_1(\omega) = t'^2 g_{nn}(\omega), \quad t_{\text{eff}}(\omega) = t'^2 g_{n0}(\omega) \tag{2}$$

depend only on the Green functions of the isolated ring. The line broadening Γ depends on the hybridisation with the leads and on their density of states, which also determines the integer $m = 1, 2$. The energy of the incoming electron is ω, and V_g is a gate voltage which may be used to control the relative energy of the ring and leads. This result is valid for systems with no spin degeneracy in the ground state.

2. Spin–Charge Separation

We consider the manifestations of SCS in an AB ring by using the t–J model as a canonical example of a strongly correlated system. The results for the transmission of the ring integrated over an energy window near the Fermi energy are shown in Fig. 2. The dips at specific flux values, $\phi_d \approx \pi(1-2n_s/N)$ for N electrons in the intermediate state, are a clear signature of SCS. These dips arise from destructive interference processes between specific pairs of states close to the Fermi energy, which then do not contribute to transport. Their dependence on the flux may be deduced by considering the $J \to 0$ limit, where the model is equivalent to the infinite–U Hubbard model and has SCS on all energy scales. Here the factorisation of the wavefunction[6,7] results in the spin part becoming equivalent only to an effective applied flux, which determines the values ϕ_d of the external flux responsible for destructive interference.[8] This nanoscopic SCS is robust for all values of the integration window, but breaks down when J is increased beyond the value t_r/L at which the spin subbands of different charge excitations become mixed. For odd N, at the flux values ϕ_d one may deduce effective spin and charge velocities which obey the relation $v_s/v_c = 1/N$.

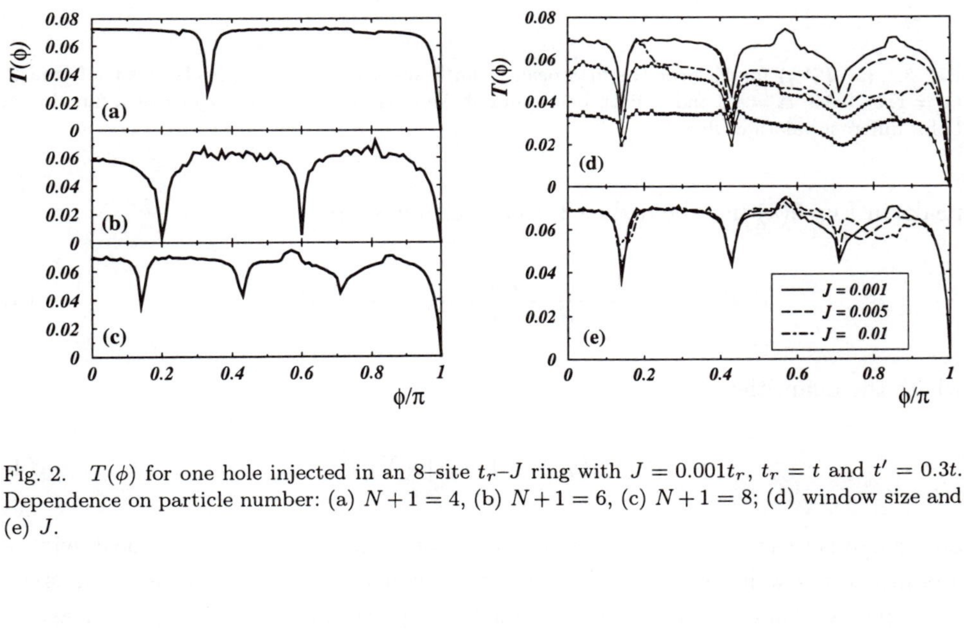

Fig. 2. $T(\phi)$ for one hole injected in an 8–site t_r–J ring with $J = 0.001t_r$, $t_r = t$ and $t' = 0.3t$. Dependence on particle number: (a) $N+1=4$, (b) $N+1=6$, (c) $N+1=8$; (d) window size and (e) J.

3. Topological Phase Transitions

To illustrate the topological charge transition we consider an IHM with $U \sim 4t$ on the AB ring. The clear disappearance at $\phi = 0$ of the integrated peak weight for $U < U_c$ [Fig. 3(a)] marks the transition between the band and correlated insulator phases, which have different ground–state parities. This transition is topological in the sense that it is detected by a topological quantum number (the charge Berry phase) whose value changes by a discrete amount even in a finite system.

For a system with no ground–state degeneracy the properties of the spectral

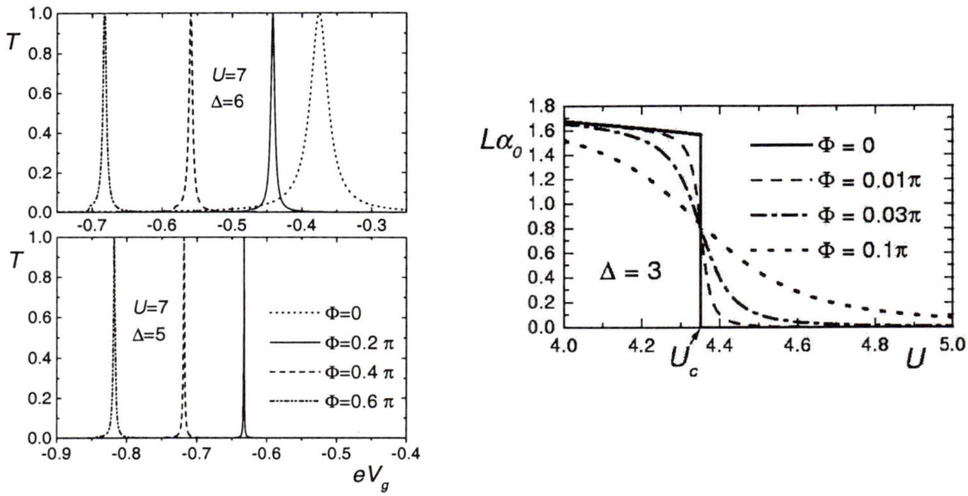

(a) (b)

Fig. 3. (a) $T(V_g)$ for V_g near the first peak in an 8–site IHM ring with leads at sites 0 and 2, $t_R = 1$, $t' = 0.2$, $\Delta = -3$ and U both below and above $U_c(L = 8) = 4.352$. (b) α_0 as a function of U for different values of flux ϕ.

peaks in Fig. 3(a) may be deduced from their lineshapes

$$T(z) \cong \frac{1}{1 + (z/w)^2} \frac{4\alpha_0\alpha_n}{(\alpha_0 + \alpha_n)^2} + O((t'/\Gamma(0))^2), \tag{3}$$

where the quantities

$$\alpha_j = |\langle g(N-1)|c_{j\sigma}|g(N)\rangle|^2 \tag{4}$$

are analgous to spectral weights for local photoemission processes. The peak widths and integrated weights depend directly on α_0, which is difficult to measure in spectroscopic experiments, but can be obtained to high resolution by transport measurements. α_0 may then be used as the order parameter for the topological transition, which occurs for all values of t: in the band insulator it increases for decreasing ϕ, whereas in the correlated insulator $\alpha_0 \to 0$ as $\phi \to 0$. At $\phi = 0$, α_0 vanishes discontinuously at U_c [Fig. 3(b)], while the transition is smoother at finite flux (parity–breaking). For a system with a spin–degenerate ground state a Kondo resonance arises, which causes spectral peaks to become plateaux. This situation, which we investigate elsewhere,[9,10] is suppressed by finite temperatures or strong magnetic fields applied transverse to the ring.

4. Summary

We have demonstrated how the AB ring may be used as a probe for nanoscopic SCS and topological phase transitions. The ring transmittance is determined by interference effects, and allows access to the spectroscopic properties of the system. The signature of SCS is a sensitive dependence of integrated transmission on flux and doping, while the topological transition is the consequence of a change of ground-state symmetry. Experimentally, one may expect to detect these effects in quantum-dot rings if these could be fabricated with sufficiently high symmetry. Annular molecules may also offer a means of detecting the charge transition, which occurs as $\phi \to 0$, but nanoscopic SCS, which requires a threading flux on the order of a single flux quantum, appears beyond reach in single molecules. We comment that if a charged wire can be made to thread the ring, which may be realised with nested nanotubes, the possibility exists of detecting also an Aharonov–Casher effect.[11]

References

1. H. J. Schulz, *Int. J. Mod. Phys.* **B5**, 57 (1991).
2. C. Kim *et al.*, *Phys. Rev. Lett.* **77**, 4054 (1996).
3. E. A. Jagla, K. Hallberg, and C. A. Balseiro, *Phys. Rev.* **B47**, 5849 (1993).
4. E. A. Jagla and C. A. Balseiro, *Phys. Rev. Lett.* **70**, 639 (1993).
5. S. R. Manmana, V. Meden, R. M. Noack and K. Schoenhammer, *Phys. Rev.* **B70**, 155115 (2004).
6. M. Ogata and H. Shiba, *Phys. Rev.* **B41**, 2326 (1990).
7. A. Schadschneider, *Phys. Rev.* **B51**, 10386 (1995).
8. K. Hallberg, A. A. Aligia, A. P. Kampf and B. Normand, *Phys. Rev. Lett.* **93**, 067203 (2004).
9. A. A. Aligia, K. Hallberg, B. Normand and A. P. Kampf, *Phys. Rev. Lett.* **93**, 076801 (2004).
10. A. A. Aligia and L. A. Salguero, *Phys. Rev.* **B70**, 075307 (2004).
11. A. Balatsky and B. Altshuler, *Phys. Rev. Lett.* **70**, 1678 (1993).

Quantum Fluids and Solids

TWO–PARTICLE–TWO–HOLE EXCITATIONS IN ^{3}He

E. KROTSCHECK, H. M. BÖHM and K. SCHÖRKHUBER
Institute for Theoretical Physics, Johannes Kepler University,
A-4040 LINZ, Austria
Eckhard.Krotscheck@jku.at

Received 1 December 2004

We describe the development of a systematic theory of excitations in strongly interacting Fermi systems. Technically, we derive the equations of motion for multi–pair excitations from a stationarity principle. This method has, in Fermi systems, so far been developed only to the level of one–particle–one–hole excitations, where it leads to the (correlated) random phase approximation (RPA). We extend the analysis here to pair excitations. Our work is motivated by the fact that time–dependent *pair* correlations are necessary for explaining the physics of the phonon–roton spectrum in ^{4}He. It is therefore plausible that the same processes also have visible effects in the excitation spectrum of ^{3}He. Further motivation is derived from recent measurements of the dynamic structure function in two–dimensional ^{3}He.

We first formulate the theory for a second quantized, weakly interacting Hamiltonian and then generalize the theory to a correlated ground state. We show that the inclusion of Jastrow–Feenberg type correlations leads to prescriptions for calculating weak effective interactions from a microscopic, strongly interacting Hamiltonian.

Keywords: Dynamic structure function of ^{3}He; correlated basis functions; pair excitations; effective interactions; density response function.

1. Introduction

The equations–of–motion method has been used as a vehicle to derive the time–dependent Hartree–Fock theory,[1,2] its analog for strongly interacting systems[3,4] as well as equations of motion for single– and multi–particle correlations in both homogeneous and inhomogeneous, strongly interacting Bose liquids. The simplest type of processes are low–lying excitations that are characterized by the quantum numbers of a single particle. This is the core idea of Landau's quasiparticle picture[5,6] and of Feynman's theory of excitations in the helium liquids.[7] In Bose systems, these excitations are simply sound waves, whereas one can have particle–hole excitations as well as sound waves in a Fermi liquid.

It was already realized by Feynman[8] that this concept is insufficient to describe higher–lying excitations, in particular the "roton" excitation in ^{4}He. An intuitively appealing solution of the problem is the introduction of Feynman–Cohen "back-flow" correlations.[8] These turn out to be a special case of "pair" excitations, *i.e.* arising from *two* particles. Such a theory has been systematically developed for Bose liquids by Jackson, Feenberg, Campbell, and collaborators.[9,10,11,12] More recently,

Apaja and Saarela[13] have carried out a very complete evaluation of the multi-particle–correlation functions appearing in the theory and arrived at an almost perfect agreement between theory and experiment up to and beyond the roton region. The appealing feature of the theory is that it is systematic in the sense that it preserves the two first energy–weighted sum rules independent of the specifics of its implementation.

2. Motivation: $S(q,\omega)$ in Three and Two Dimensions

Let us first look at the "conventional" way of discussing the density–density response function in a Fermi liquid: One assumes an RPA form

$$\chi^{(\mathrm{RPA})}(q,\omega) = \frac{\chi_0(q,\omega)}{1 - \widetilde{V}_{\text{p-h}}(q)\,\chi_0(q,\omega)} \tag{1}$$

where $\chi_0(q,\omega)$ is the Lindhard function, and $\widetilde{V}_{\text{p-h}}(q)$ a suitably chosen effective interaction or "polarization–potential". One way to choose this interaction is to define it through the two sum rules

$$S(q) = \int d\omega\, S(q,\omega) \equiv \frac{\hbar}{\pi}\int d\omega\, \Im m\,\chi(q,\omega), \qquad \frac{\hbar^2 q^2}{2m} = \frac{\hbar^2}{\pi}\int d\omega\,\omega\,\Im m\,\chi(q,\omega). \tag{2}$$

The dynamic structure function of three–dimensional ^{3}He at saturation density obtained in this manner is shown in Fig. 1.

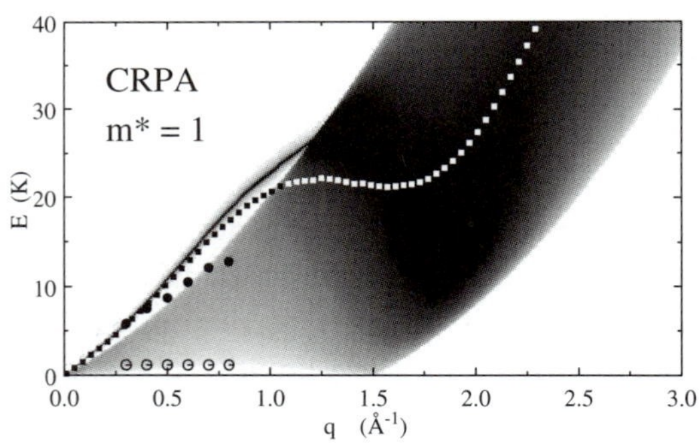

Fig. 1. The figure shows the dynamic structure function $S(k,E)$ for ^{3}He at saturation density, obtained as described in the text. Darker shades correspond to higher values of $S(k,E)$. Also shown are experimental values for the phonon (dark, filled dots) and magnon (open dots) contributions from Ref. 14 and the Feynman dispersion relation (open and filled squares).

We observe the familiar problem that the collective density–density fluctuation mode is energetically much higher than found in experiments. A popular cure for the problem is to introduce an average effective mass in the Lindhard function. With a reasonable effective mass, relatively good agreement with experiments can be obtained. We believe, however, that such a procedure is unsatisfactory: introducing an effective mass violates the sum rules (2). However, the left hand sides of Eqs. (2) are known either exactly, or with very high accuracy. These sum rules can only be re–established by introducing further physical effects like, for example, a specific energy dependence of the particle–hole interaction whose specific analytic form is hard to justify.

The case of two–dimensional ^{3}He is an even more striking example that a simple modification of the RPA form by introducing an effective mass is an improper way to resolve the discrepancy. Recent measurements[15] allow for the identification of a collective mode, but this mode seems to be inside a continuum of excitations (Fig. 2, left part). Introducing an effective mass in a simple RPA calculation, would, however, lower *both* the particle–hole continuum and the collective mode (Fig. 2, right part) but it can not move the collective mode into the continuum.

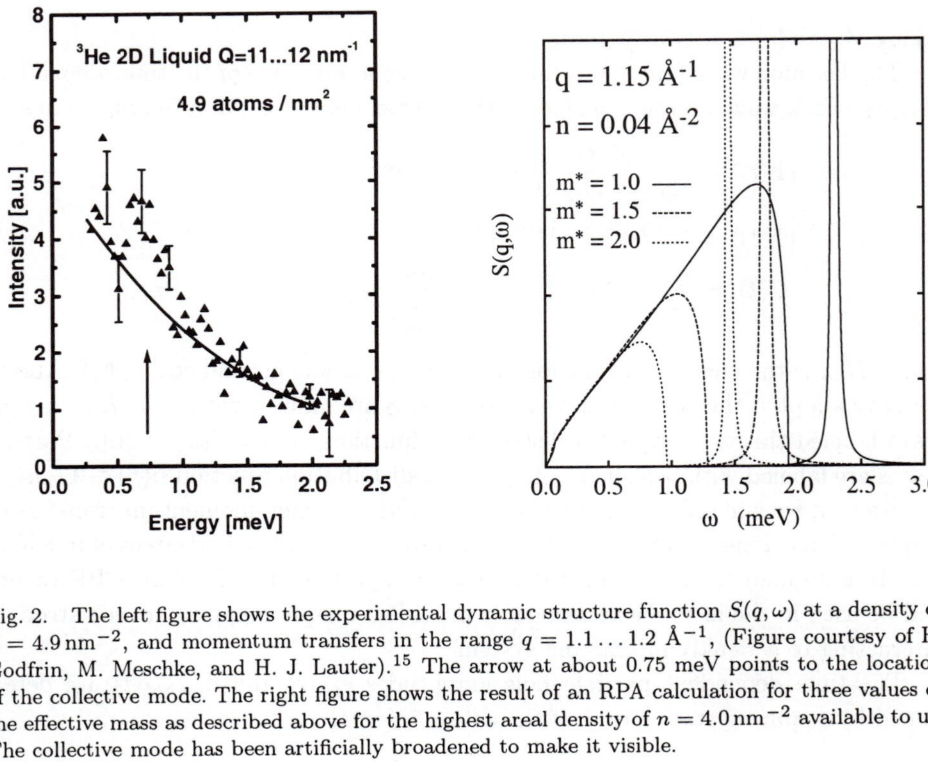

Fig. 2. The left figure shows the experimental dynamic structure function $S(q,\omega)$ at a density of $n = 4.9\,\mathrm{nm}^{-2}$, and momentum transfers in the range $q = 1.1 \ldots 1.2$ Å^{-1}, (Figure courtesy of H. Godfrin, M. Meschke, and H. J. Lauter).[15] The arrow at about 0.75 meV points to the location of the collective mode. The right figure shows the result of an RPA calculation for three values of the effective mass as described above for the highest areal density of $n = 4.0\,\mathrm{nm}^{-2}$ available to us. The collective mode has been artificially broadened to make it visible.

From a physical point of wiew, further strong evidence that introducing an effective mass in the Lindhard function does not describe the right physics in ^{3}He comes from ^{4}He: In Bose systems, the procedure of using the RPA together with the sum rules (2) for defining an effective interaction leads to the famous Feynman dispersion relation $\omega(q) = \hbar^2 q^2/2mS(q)$. Again, at wavelengths in the regime of the roton minimum this energy is too high by about a factor of two. The problem in Bose liquids is naturally solved by introducing fluctuations of the wave function at a wavelength comparable to the interparticle distance. Since ^{3}He and ^{4}He are very similar systems with excitations at comparable wavelengths, it appears therefore appropriate to improve the theoretical description of excitations in ^{3}He also by introducing time–dependent pair correlations.

3. Weakly Interacting Fermions

In order to examine the anticipated structure of the resulting equations, we start with the model of a weakly interacting system. We assume a Hamiltonian in second quantized form

$$H = \sum_{\alpha} t_\alpha a^\dagger_\alpha a_\alpha + \frac{1}{2} \sum_{\alpha\beta\gamma\delta} \langle \alpha\beta | V | \delta\gamma \rangle \, a^\dagger_\alpha a^\dagger_\beta a_\gamma a_\delta \,, \tag{3}$$

where $V = V(r)$ is a *local* potential.

The Fermion wave function is taken to be a generalization of the time–dependent Hartree–Fock wave function *including* time–dependent two–particle amplitudes:

$$\begin{aligned} |\Psi(t)\rangle &= \frac{|\Phi(t)\rangle}{\langle \Phi(t) \mid \Phi(t)\rangle^{1/2}} \equiv \mathrm{e}^{-iH_{\mathbf{oo}}\, t/\hbar} \, |\Psi_0(t)\rangle \,, \\ |\Phi(t)\rangle &= \mathrm{e}^{-iH_{\mathbf{oo}}\, t/\hbar} \, e^{\frac{1}{2}U(t)} \, |\Phi_0\rangle \,, \\ U(t) &= \sum_{ph} c_{ph}(t) \, a^\dagger_p a_h + \frac{1}{2} \sum_{pp'hh'} d_{pp'hh'}(t) \, a^\dagger_p a^\dagger_{p'} a_{h'} a_h \,, \end{aligned} \tag{4}$$

where $H_{\mathbf{oo}}$ is the energy expectation value for the model ground state $|\Phi_0\rangle$. Above, we have adopted the convention that hole states are labeled with $h, h', h_i, \ldots$, each state labeled this way implies a distribution function $n(h) = \theta(k_F - |\mathbf{h}|)$. Particle states are labeled with $p, p', p_i, \ldots$, implying a distribution function $\bar{n}(p) \equiv 1 - n(p)$. For brevity, we will only spell out Fermi distributions when momentum transfers or combinations of momenta are involved. We also suppress all spin degrees of freedom. The Hamiltonian (3) is interpreted as a *weak, effective* Hamiltonian; CBF theory will be used in the in the next section to provide a unique mapping from a strongly interacting to a weakly interacting system.

The time–dependent particle–hole amplitudes $c_{ph}(t)$ and $d_{pp'hh'}(t)$ are determined by requiring stationarity of the action integral

$$\mathcal{S}\left[c_{ph}, c^*_{ph}, d_{pp'hh'}, d^*_{pp'hh'}\right] = \int dt \, \mathcal{L}(t) \,, \tag{5}$$

where the Lagrangian has the form

$$\begin{aligned} \mathcal{L}(t) &= \langle \Psi(t) | \; H + H_{\mathrm{ext}}(t) - i\hbar\frac{\partial}{\partial t} \; |\Psi(t)\rangle \\ &= \langle \Psi_0(t) | \; H' + H_{\mathrm{ext}}(t) - i\hbar\frac{\partial}{\partial t} \; |\Psi_0(t)\rangle \end{aligned} \tag{6}$$

with $H' = H - H_{\mathbf{oo}}$.

The term $H_{\rm ext}(t) = \sum_i h_{\rm ext}(\mathbf{r}_i;t)$ in Eq. (6) describes a small, external perturbation which is real and has, without loss of generality, harmonic time dependence:

$$h_{\rm ext}(\mathbf{r},t) = h_{\rm ext}(\mathbf{r},\omega)\ \left(e^{i\omega t} + e^{-i\omega t}\right)\ . \tag{7}$$

We can then also assume harmonic time–dependence of the particle–hole amplitudes.

The next task is to evaluate the action integral to second order in the amplitudes c_{ph} and $d_{pp'hh'}$, treating the perturbing external field $H_{\rm ext}(t)$ as a first order quantity. We are mainly interested in the density fluctuation $\delta\rho(\mathbf{r};t)$ induced by this external field. To first order in the fluctuations excitation operator $U(t)$, we obtain for the expectation value of the density operator $\hat\rho(\mathbf{r})$

$$\delta\rho(\mathbf{r};t) = \langle\Psi(t)|\,\hat\rho(\mathbf{r})\,|\Psi(t)\rangle \approx \frac{1}{2}\sum_{ph}\left[c^*_{ph}(t)\ \phi^*_p(\mathbf{r})\phi_h(\mathbf{r}) + c_{ph}(t)\ \phi^*_h(\mathbf{r})\phi_p(\mathbf{r})\right]\ , \tag{8}$$

and of $H_{\rm ext}$:

$$\langle\Psi(t)|\,H_{\rm ext}(t)\,|\Psi(t)\rangle \approx \frac{1}{2}\sum_{ph}\left[c^*_{ph}(t)\ h_{ph}(t) + c_{ph}(t)\ h^*_{ph}(t)\right]\ . \tag{9}$$

For the time–derivative term, we obtain

$$\langle\Psi_0(t)| - i\hbar\frac{\partial}{\partial t}\,|\Psi_0(t)\rangle = \frac{\hbar}{4}\Im m\left[\sum_{ph} c^*_{ph}(t)\dot c_{ph}(t) + \frac{1}{2}\sum_{pp'hh'} d^*_{pp'hh'}(t)\dot d_{pp'\{hh'\}_a}(t)\right]\ . \tag{10}$$

The notation $\{hh'\}_a$ indicates that the labels h and h' should be antisymmetrized. Above, we have used the fact that terms that can be written as a total time–derivative can be omitted.

The remaining "Hamiltonian" term has the general form

$$\langle\Psi(t)|\,H'\,|\Psi(t)\rangle \equiv \delta H_{cc}(t) + \delta H_{cd}(t) + \delta H_{dc}(t) + \delta H_{dd}(t)\ , \tag{11}$$

where contributions from single pair amplitudes and two pair amplitudes are marked by c and d, respectively.

To simplify the calculation, we keep only those terms that would (a) also be included in the derivation of the random phase approximation, and (b) are also kept in the boson version of the theory. The first approximation implies the omission of exchange terms. We also restrict the pair excitation operator to *local* correlations. This means that we write the pair excitations in terms of *density fluctuations*

$$\sum_{pp'hh'} d_{pp'hh'}\ a^\dagger_p a^\dagger_{p'} a_{h'} a_h \quad\longrightarrow\quad \sum_{qq'} d(\mathbf{q},\mathbf{q}')\left[\hat\rho_{\mathbf{q}}\hat\rho_{\mathbf{q}'} - \hat\rho_{\mathbf{q}+\mathbf{q}'}\right]\ , \tag{12}$$

where $\hat\rho_{\mathbf{q}} = \sum_k a^\dagger_{\mathbf{k}+\mathbf{q}} a_{\mathbf{k}}$ is the density operator.

A further approximation is made to simplify the *c-d* coupling matrix elements: We replace in the Lagrangian the one–particle–one–hole amplitudes $c_{ph}(t)$ in the terms $\delta H_{12}(t)$ and $\delta H_{21}(t)$ by their Fermi–sea averages $c(\mathbf{q};t)$ defined as

$$c_{h+q,h}(t) \quad \longrightarrow \quad c(\mathbf{q};t) = N\frac{\sum_h n(h)\bar{n}(\mathbf{h}+\mathbf{q})\, c_{h+q,h}(t)}{\sum_h n(h)\bar{n}(\mathbf{h}+\mathbf{q})}\,. \tag{13}$$

We shall skip the rather complicated technical manipulations[16] and proceed to discuss the final results. We need the partial Lindhard functions

$$\chi_0^{(\pm)}(q,\omega) = \pm\frac{1}{N}\sum_h \frac{n(h)\,\bar{n}(\mathbf{h}+\mathbf{q})}{\hbar\omega \mp e_{h+q\,h} + i\eta}\,, \tag{14}$$

where $e_{h+q\,h} \equiv e_{h+q} - e_h$ are differences of particle–and hole energies. $\chi_0^{(+)}(q,\omega)$ and $\chi_0^{(-)}(q,\omega)$ add up to the full Lindhard function

$$\chi_0(q,\omega) = \chi_0^{(+)}(q,\omega) + \chi_0^{(-)}(q,\omega)\,. \tag{15}$$

In terms of these quantities, from the approximations outlined above, one can derive a closed–form expression for the density–density response function of a weakly interacting system, that has a a rather plausible form:

$$\chi(q,\omega) = \frac{\kappa(q,\omega)}{1-\kappa(q,\omega)\widetilde{V}(q)} \tag{16}$$

where

$$\kappa(q,\omega) = \frac{\chi_0^{(+)}(q,\omega)}{1-\chi_0^{(+)}(q,\omega)\mathcal{W}(q,-\omega)} + \frac{\chi_0^{(-)}(q,\omega)}{1-\chi_0^{(-)}(q,\omega)\mathcal{W}(q,\omega)}\,, \tag{17}$$

$$\mathcal{W}(q,\omega) = \frac{1}{2}\sum_{\mathbf{q}'} \frac{\overline{\mathcal{W}}_3(\mathbf{q}-\mathbf{q}',\mathbf{q}')\overline{\mathcal{W}}_3(\mathbf{q}-\mathbf{q}',\mathbf{q}')}{\hbar\omega + \varepsilon(\mathbf{q}-\mathbf{q}') + \varepsilon(q')} \tag{18}$$

where

$$\overline{\mathcal{W}}_3(\mathbf{q},\mathbf{q}') = \frac{\tilde{V}(q)\,S_F(q) + \tilde{V}(q')\,S_F(q')}{\sqrt{S_F(q)\,S_F(q')}}\,\frac{S_F^{(3)}(\mathbf{q},\mathbf{q}',\mathbf{q}+\mathbf{q}')}{S_F(\mathbf{q}+\mathbf{q}')} \tag{19}$$

is a three-"phonon" vertex, and $\varepsilon(q)$ is the weakly–interacting approximation of the Bogoljubov spectrum

$$\varepsilon(q) = \frac{\hbar^2 q^2}{2mS_F(q)} + S_F(q)\widetilde{V}(q) \approx \frac{\hbar^2 q^2}{2mS_F(q)}\sqrt{1+\frac{4mS_F^2(q)\widetilde{V}(q)}{\hbar^2 q^2}}\,. \tag{20}$$

Finally, $S_F(q)$ and $S_F^{(3)}(\mathbf{q},\mathbf{q}',\mathbf{q}+\mathbf{q}')$ are the static two– and three–body structure functions of the non–interactinf Fermi system.

A theory of the kind outlined here is not *directly* applicable for strongly interacting systems; one should re–interpret the interaction in the Hamiltonian as an *effective* interaction. Then, there is of course no reason that this interaction should be the same in all channels. An unambiguous method for obtaining such interactions from an underlying microscopic Hamiltonian is provided by CBF theory, which we will describe in the next section.

4. Strongly Interacting Fermions

Correlated basis functions (CBF) theory is often considered to be mainly a theory for the ground state of strongly interacting Fermions. For that purpose, the ground state wave function is written in the form

$$|\Psi_0(1,\ldots,N)\rangle = \frac{F(1,\ldots,N)\,|\Phi_{\mathbf{o}}(1,\ldots,N)\rangle}{\langle\Phi_{\mathbf{o}}|F^\dagger F|\Phi_{\mathbf{o}}\rangle^{1/2}}\,. \tag{21}$$

where $\Phi_{\mathbf{o}}(1,\ldots,N)$ is a model wave function, normally a Slater–determinant of plane waves, and $F(1,\ldots,N)$ is the correlation operator, suitably chosen and optimized to model the important features of the interacting system. The most practical — and most successful — form of the correlation operator is the Jastrow–Feenberg form[17]

$$F(1,\ldots,N) = \exp\frac{1}{2}\left[\sum_{1\le i<j\le N} u_2(\mathbf{r}_i,\mathbf{r}_j) + \sum_{1\le i<j<k\le N} u_3(\mathbf{r}_i,\mathbf{r}_j,\mathbf{r}_k) + \ldots\right]\,. \tag{22}$$

CBF theory is, however, not limited to ground state correlations, and its application to excited states provides the aforementioned vehicle for dealing with strong interactions and deriving effective interactions. To do so, CBF theory generates a complete set of correlated and normalized N-particle basis states from the excited state Slater determinants $|\Phi_{\mathbf{m}}\rangle$ through

$$|\mathbf{m}\rangle = \frac{F\,|\Phi_{\mathbf{m}}\rangle}{\langle\Phi_{\mathbf{m}}|F^\dagger F|\Phi_{\mathbf{m}}\rangle^{1/2}}\,. \tag{23}$$

Even though the states $\{|\mathbf{m}\rangle\}$ are not orthogonal, perturbation theory and the equations of motions outlined above can be formulated in terms of these states. The generalization of the wave function (4) for excited states to such a strongly correlated basis set is

$$|\Psi(t)\rangle = \frac{F\;|\Phi(t)\rangle}{\langle\Phi(t)|\;F^\dagger F\;|\Phi(t)\rangle^{1/2}} \equiv \mathrm{e}^{-iH_{\mathbf{oo}}t/\hbar}\,|\Psi_0(t)\rangle \tag{24}$$

where $|\Phi(t)\rangle$ is given in Eq. (4), and $H_{\mathbf{oo}}$ is the expectation value of the Hamiltonian with respect to the *correlated* state (21). The particle–hole amplitudes $c_{ph}(t)$ and $d_{pp'hh'}(t)$ are again determined by stationarity of the action integral (5), the only difference being that the state $|\Psi(t)\rangle$ is now correlated.

We can then formulate, in analogy to section 3, the action integral to second order in the particle–hole amplitudes. The time–derivative term yields

$$\left\langle \Psi_0(t)\left|-i\hbar\frac{\partial}{\partial t}\right|\Psi_0(t)\right\rangle = \frac{\hbar}{4}\Im m\frac{\langle\Phi_0|\,U^\dagger(t)F^\dagger F\dot{U}(t)\,|\Phi_0\rangle}{\langle\Phi_0|\,F^\dagger F\,|\Phi_0\rangle}\,. \tag{25}$$

The second–order Lagrangian is then

$$\begin{aligned}\mathcal{L}^{(2)} &= \frac{1}{8}\frac{\langle\Phi_0|\,U^{\dagger 2}(t)F^\dagger H'F + 2U^\dagger(t)F^\dagger H'FU(t) + F^\dagger H'FU^2(t)\,|\Phi_0\rangle}{\langle\Phi_0|\,F^\dagger F\,|\Phi_0\rangle}\\ &+ \frac{\hbar}{4}\Im m\frac{\langle\Phi_0|\,U^\dagger(t)F^\dagger F\dot{U}(t)\,|\Phi_0\rangle}{\langle\Phi_0|\,F^\dagger F\,|\Phi_0\rangle}\,.\end{aligned} \tag{26}$$

Dealing with the full particle–hole structure of the pair excitations is a rather tedious task leading to matrix–elements of four–body operators. In section 3 progress was made by restricting the pair excitations to pair–density fluctuations, and by replacing the particle–hole amplitudes in the 1-2 coupling term by their Fermi–sea averages (13). Both of these assumptions imply significant simplifications because we can in all those quantities that contain one or two pair excitation operators use the formal results of the boson theory. We must, of course, use the cumulant expansions of the distribution functions appropriate for Fermi statistics.

The remaining manipulations are nevertheless quite lengthy, and we proceed here, as above, to simply displaying the final result. One arrives at a density–density response function of the form

$$\chi^{(\mathrm{CBF})}(q,\omega) = \frac{\kappa(q,\omega)}{1-\kappa(q,\omega)\widetilde{V}_{\text{p-h}}(q) + V_{\mathrm{mp}}(q,\omega)} \tag{27}$$

where $\kappa(q,\omega)$ has the same formal form as in Eq. (18). Here, in particular the quantity $\widetilde{V}_{\text{p-h}}(q)$ is the same quantity that is known from the correlated RPA theory[4] and from the optimization of the Fermi problem. The three-phonon term $\mathcal{W}(q,\omega)$ is given in Eq. (18), but now the spectrum is the full Bogoljubov spectrum,

$$\varepsilon(q) = \frac{\hbar^2 q^2}{2mS_F(q)}\sqrt{1+\frac{4mS_F^2(q)\widetilde{V}_{\text{p-h}}(q)}{\hbar^2 q^2}}\,. \tag{28}$$

All further interactions entering the expression for $\mathcal{W}(q,\omega)$ are also given by the microscopic theory. The only additional term, in which the response function of the fully correlated theory deviates from the formal structure of the "weakly interacting" theory is the multiphonon term $V_{\mathrm{mp}}(q,\omega)$. This term is rather complicated and has no immediately transparent structure, we therefore will not give its analytic form here.

To conclude this section, we return to the discussion of sum rules in Sec. 2. In complete analogy to Bose theory [11] it is easy to prove that

$$\begin{aligned}\frac{1}{\pi}\int d\omega\, \Im m\, \chi^{(\mathrm{RPA})}(q,\omega) &= \frac{1}{\pi}\int d\omega\, \Im m\, \chi^{(\mathrm{CBF})}(q,\omega)\\ \frac{1}{\pi}\int d\omega\, \omega\, \Im m\, \chi^{(\mathrm{RPA})}(q,\omega) &= \frac{1}{\pi}\int d\omega\, \omega\, \Im m\, \chi^{(\mathrm{CBF})}(q,\omega)\,.\end{aligned} \tag{29}$$

Note in particular that this statement is independent of the specific form of the three–phonon coupling term. This means that the static particle–hole interaction $\widetilde{V}_{\text{p-h}}(q)$ is *uniquely* defined from the static structure function.

5. Results

We have carried out calculations for three different approximations for the density–density response function. Input to all of these calculations were ground state results for $S(k)$ obtained in Ref. 18. The simplest approximation is the "correlated RPA" which omits fluctuating pair correlations. The result obtained for ^{3}He at its equilibrium density is shown in Fig. 1 and was already discussed above.

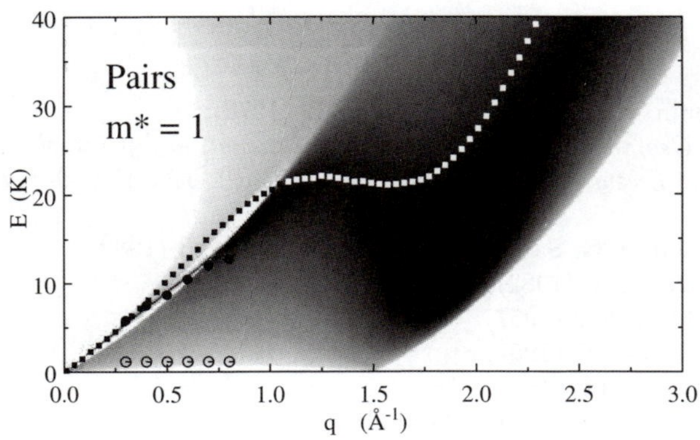

Fig. 3. Same as Fig. 1 for the theory including correlaed pair excitations, but without multiphonon correction. Note that the calculated collective mode (represeted by a dark line) agrees well with the experimentally obtained data (dark circles).

Fig. 3 shows the result of CBF theory including pair fluctuations, but omitting the multiphonon correction $V_{\rm mp}(q)$. For the effective interactions $\widetilde{V}_{\rm p\text{-}h}(q)$ and $\mathcal{W}(q,\omega)$, however, the full CBF results have been used (as outlined following Eq. (27). Obviously, the results are significantly improved, the peak of the calculated zero–sound mode is, in the region where it is not Landau damped, practically identical with the experimental values. As we see, a modification of the effective mass is unnecessary also in ^{3}He.

Finally, we mention that the inclusion of the multiphonon correction term $V_{\rm mp}(q,\omega)$ provides little additional improvement,[16] it has therefore been omitted here.

6. Summary

The work described in this paper is, obviously, a beginning. After the very successful introduction of dynamic multi–particle correlations in ^{4}He, its extension to ^{3}He is a logical step that has been overdue for perhaps 30 years. As in ^{4}He, the inclusion of correlated pair excitations is necessary for obtaining the correct phonon dispersion also in ^{3}He.

Evidently, many technical and physical questions remain open.

Technically, one would like to see if some of the approximations, specifically the "collective" approximation of the 2-2-coupling term, can be relaxed. This might also permit a more direct interpretation of the theory in terms of Feynman diagrams.

Physically, the most prominent question is the inclusion of spin fluctuations. These are important low–low–lying modes; their inclusion is needed for understanding the origin of the large effective mass in ^{3}He. Work in this direction is in progress.

Acknowledgments

This work was supported, in part, by the Austrian Science Fund (FWF) under project P-12832-TPH. We also thank H. Godfrin for giving us access to the data of Ref. 15 prior to publication.

References

1. A. K. Kerman and S. E. Koonin, Ann. Phys. (NY) **100**, 332 (1976).
2. P. Kramer and M. Saraceno, *Geometry of the time-dependent variational principle in quantum mechanics*, Vol. 140 of *Lecture Notes in Physics* (Springer, Berlin, Heidelberg, and New York, 1981).
3. J. M. C. Chen, J. W. Clark, and D. G. Sandler, Z. Physik A **305**, 223 (1982).
4. E. Krotscheck, Phys. Rev. A **26**, 3536 (1982).
5. L. D. Landau, Sov. Phys. JETP **3**, 920 (1957).
6. L. D. Landau, Sov. Phys. JETP **5**, 101 (1957).
7. R. P. Feynman, Phys. Rev. **94**, 262 (1954).
8. R. P. Feynman and M. Cohen, Phys. Rev. **102**, 1189 (1956).
9. H. W. Jackson and E. Feenberg, Ann. Phys. (NY) **15**, 266 (1961).
10. H. W. Jackson, Phys. Rev. A **8**, 1529 (1973).
11. H. W. Jackson, Phys. Rev. A **9**, 964 (1974).
12. C. C. Chang and C. E. Campbell, Phys. Rev. B **13**, 3779 (1976).
13. V. Apaja and M. Saarela, Phys. Rev. B **57**, 5358 (1998).
14. H. R. Glyde *et al.*, Phys. Rev. B **61**, 1421 (2000).
15. H. Godfrin, M. Meschke, and H. J. Lauter, to be published.
16. K. E. Schörkhuber, Ph.D. thesis, Institut für Theoretische Physik, Johannes Kepler Universität Linz, Austria, 2002.
17. E. Feenberg, *Theory of Quantum Fluids* (Academic, New York, 1969).
18. E. Krotscheck, J. Low Temp. Phys. **119**, 103 (2000).

MONOLAYER CHARGED QUANTUM FILMS: A QUANTUM SIMULATION STUDY

KEOLA WIERSCHEM

Physics Department, Florida State University, Tallahassee, Florida, 32306-4350
keola@martech.fsu.edu

EFSTRATIOS MANOUSAKIS

Physics Department, Florida State University, Tallahassee, Florida, 32306-4350
stratos@martech.fsu.edu

Received 1 December 2004

We use path-integral Monte Carlo (PIMC) to study the effects of adding a long-range repulsive Coulomb interaction to the usual Van der Waals interaction between two atoms of a submonolayer quantum film such as helium on graphite or a pure two-dimensional superfluid. Such interactions frustrate or compete with the natural tendency of the system for phase separation namely to form a macroscopic liquid or solid phase. We find that as a function of the relative strength of the long-range repulsion, surface coverage and temperature, the system undergoes a series of transformations, including a triangular Wigner-like crystal of clusters, a charge stripe-ordered phase and a fluid phase. The goal of these studies is to understand the role of quantum fluctuations when such competing interactions appear together with formation of preexisting electron pairs as might be the case in cuprate superconductors.

Keywords: Superfluid films; stripes phases; charged bosons; cuprate superconductors.

1. Introduction

In a number of physical systems where there are competing interactions such as a short-range attractive part and a long-range repulsion, a rich phase diagram is expected. In the particular case where the long-range repulsion is the usual Coulomb interaction, nuclear matter and electrons in a solid such as the cuprate superconductors are examples of this case. In the latter case, the electrons in the material can find themselves experiencing an attractive interaction which might also lead to pairing or phase separation, but when they form clusters the accumulated charge cannot exceed a threshold value due to the Coulomb repulsion.

In this work we focus our attention to the case of a two dimensional quantum fluid of charged bosons with short-range attractive interaction and a hard core repulsion. This model, for example, could be used to understand certain aspects of the conduction electrons of the copper-oxide plane of the cuprate superconductors. It has been found[1,2,3] that in cuprate superconductors, as modeled by the so-called $t-J$ model, holes tend to phase separate in a hole rich and an antiferromagnetically ordered region. A pair of holes in a quantum antiferromagnet is strongly bound[4] and in fact the hole rich phase can be understood as a fluid of weakly attracted bosons–the bound pairs of holes.[5] In this paper, we consider such a system of bosons which interact via a short range attractive interaction, a hard core repulsion and a long-range Coulomb repulsion. A system of interacting bosons is free from the infamous minus problem which hinders the quantum simulation of fermions. Therefore we should be able to address the phase diagram of this model without serious difficulties.

In the present paper our goal is to understand the influence of a weak Coulomb repulsion on a quantum Van der Waals-Bose liquid monolayer film. Our model of the neutral Van der Waals-Bose liquid is a liquid helium film, because we have gained understanding of the phases of this system from experimental studies[6,7,8] and from our previous PIMC studies.[9] More importantly, using the PIMC method most of the observed features of these films and the phase diagram of each layer as a function of surface coverage can be reproduced with detail. In these PIMC studies [9] of submonolayer, second, third and higher layers of helium on graphite well-known helium-helium potential and helium-graphite interaction were used without the introduction of any fitting parameter. In this paper, we begin from the Hamiltonian that describes a submonolayer helium film on graphite and we modify it by adding a weak Coulomb repulsion term to the helium-helium interaction potential which depends on a coupling constant α. We will use PIMC method to study such a film as a function of temperature, surface density and the parameter α.

2. The Approach

For particle-substrate interaction we used the laterally averaged helium-graphite interaction[10] used in previous PIMC studies of helium adsorption on graphite.[9] In order to study the role of a long-range Coulomb repulsion on the Van der Waals helium-helium atom interaction we added a $1/r$ term to a Lennard-Jones interaction potential as follows:

$$V(r) = -4\epsilon\Big(\big(\frac{\sigma}{r}\big)^6 - \big(\frac{\sigma}{r}\big)^{12}\Big) + \alpha V_C(L,r) \tag{1}$$

where $V_C(L,r)$ is the $1/r$ modified to include the influence of the periodic image charges in a finite-size $L \times L$ lattice with periodic boundary conditions. This is implemented for two-dimensions by the well-known Ewald summation method.[12] For any given value of the parameter α we adjust the value of ϵ in order for the minimum of the Lennard-Jones potential to be at the same value $V_{min} = 10°K$. This is the approximate minimum value of the Aziz potential[11] that describes the helium-helium interaction and it is used to study neutral helium quantum films. The total interaction potential is shown in Fig. 1 for various values of the parameter α. Notice that in order to keep V_{min} a constant, increasing the value of α requires an increased value of ϵ and this leads to a harder core potential and a more attractive force dragging the neighbor to toward the potential minimum.

We will use the path-integral Monte Carlo method (PIMC) for superfluid films as developed in Ref. 9. We have carried out simulations at $T = 0.5°K$ and $T = 1°K$ for all 100 combinations of 10 density values starting from $\rho = 0.0120\text{Å}^{-2}$ up to 0.0480Å^{-2} with 10 values of α starting from $\alpha = 50°K\text{Å}$ up to $\alpha = 500°K\text{Å}$ in increments of $50°K\text{Å}$. We kept the number of particles fixed at 30 and we varied the size of the simulation cell in order to vary the density. We have computed the pair distribution function, and contour plots of the particle probability density for all of the above parameter values. Approximately 10,000 Monte Carlo sweeps were used for thermalization and about 10,000 sweeps for measurements of observable.

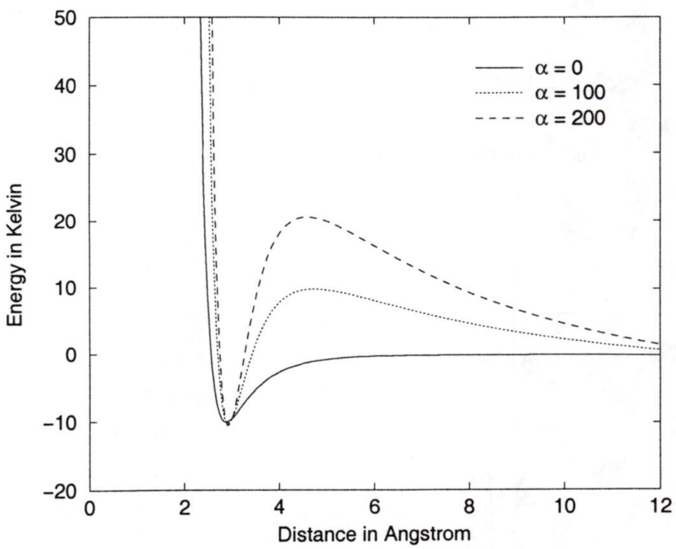

Fig. 1. The total interaction potential is shown for various values of the parameter α. The value of ϵ is adjusted such that the minimum of the interaction is fixed at $-10°K$.

3. Results

In Fig. 2 we give a thermalized space-time configuration of the system, contour plots of the probability distribution averaged over 10,000 such configurations and the pair distribution function for surface density $\rho = 0.0120\text{Å}^{-2}$ and temperature $T = 500mK$ for various values of α. We notice that while for this surface density in the absence of the Coulomb repulsion the system phase separates, the presence of even weak Coulomb interaction tends to segregate the particles into clusters which spread over the entire film surface. In addition, the pair distribution function develops a second peak at the average inter-cluster distance. There is still a peak near the minimum of the Van der Waals attractive part of the interaction but for small α this peak is much smaller compared to that of the neutral fluid. Notice that

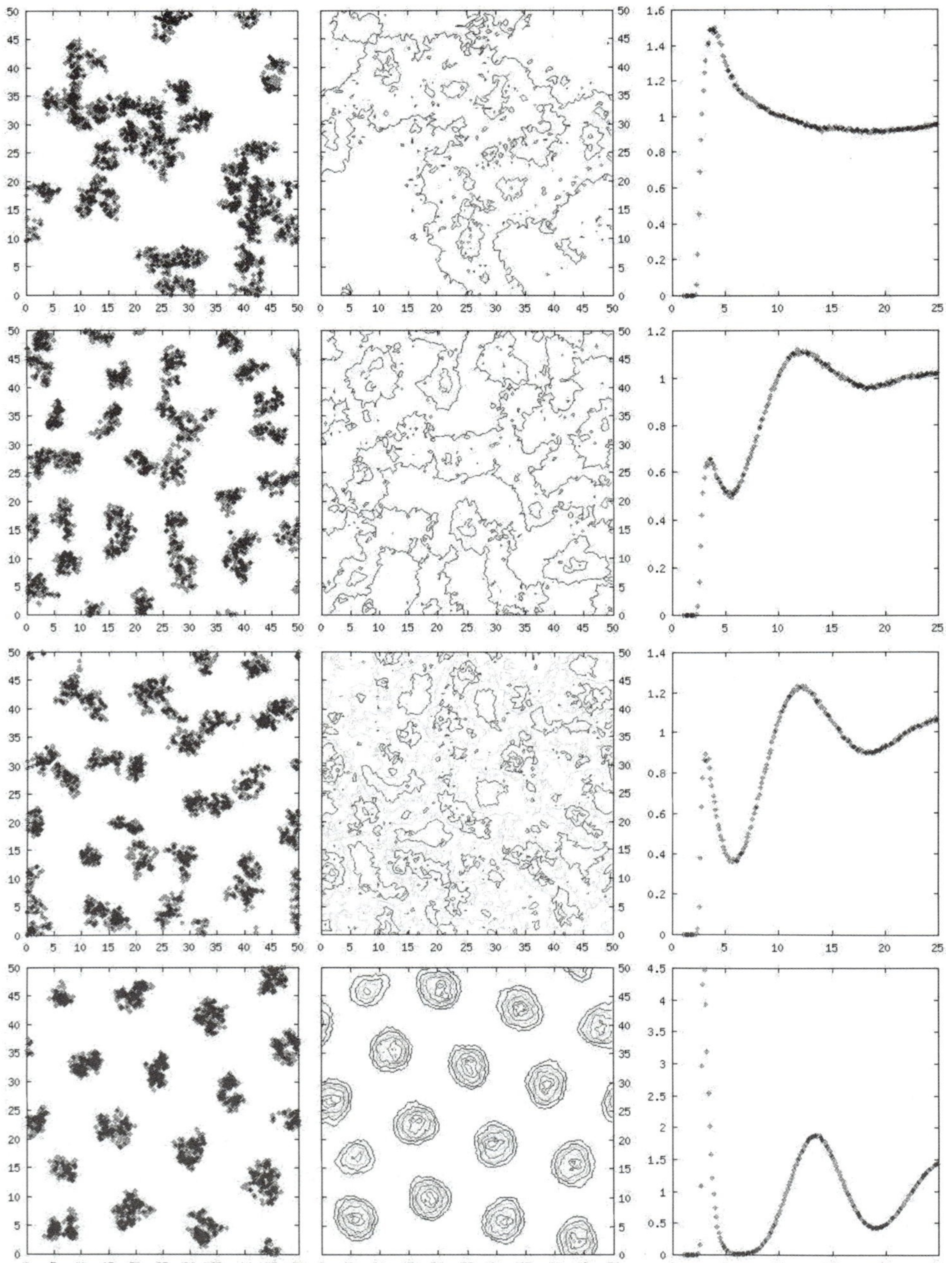

Fig. 2. A thermalized space-time configuration, the contour plot and the pair distribution at surface density 0.0120Å^{-2}, $T = 0.5°K$ and for $\alpha = 0$ (top row) $\alpha = 50°K\text{Å}$ (second row from the top) $\alpha = 100°K\text{Å}$ (third from top row) and $\alpha = 350°K\text{Å}$ (fourth row from the top)

as we raise the strength of the Coulomb repulsion the system forms a triangular superlattice of clusters. The pair distribution function has a sharp peak near the minimum of the Van der Waals attraction and an oscillatory behavior with a period equal to the distance between nearest neighbor clusters.

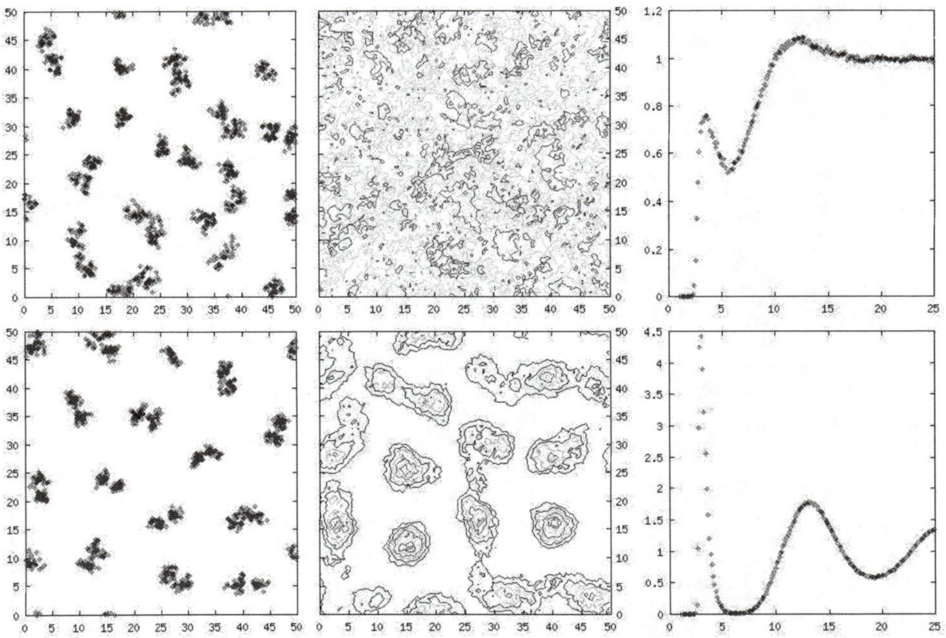

Fig. 3. A thermalized space-time configuration, the contour plot and the pair distribution at surface density $0.0120 \AA^{-2}$, $T = 1°K$ and $\alpha = 50°K\AA$ (top row) and $\alpha = 350°K\AA$ (bottom row)

By raising the temperature to $1°K$ the picture remains similar as shown in Fig. 3 with the tendency for the clusters and the particles to fluctuate more which will lead to the melting of the clusters at higher temperature. It is not clear if the system will become an ordered superlattice of such clusters or will remain a fluid at zero temperature at small values of α. It is well known that, strictly speaking, a two-dimensional system such as ours cannot break a continuous symmetry and crystallize at any finite temperature; however, a finite-size system may appear as ordered because the correlation length for low enough temperature becomes larger than the system size. We have observed that, as a function of the simulation time, a finite-size 2D system drifts as a whole. As a result, while within a relatively short time-scale the system appears ordered if we average configurations over a much larger time scale, the system will appear disordered in a contour plot. This time scale depends on the value of α. For small values of α it is easier for the system to migrate as a whole while keeping the particle correlations. This is demonstrated in the figures for $\alpha = 50, 100°K\AA$ where, while the particle correlations show that

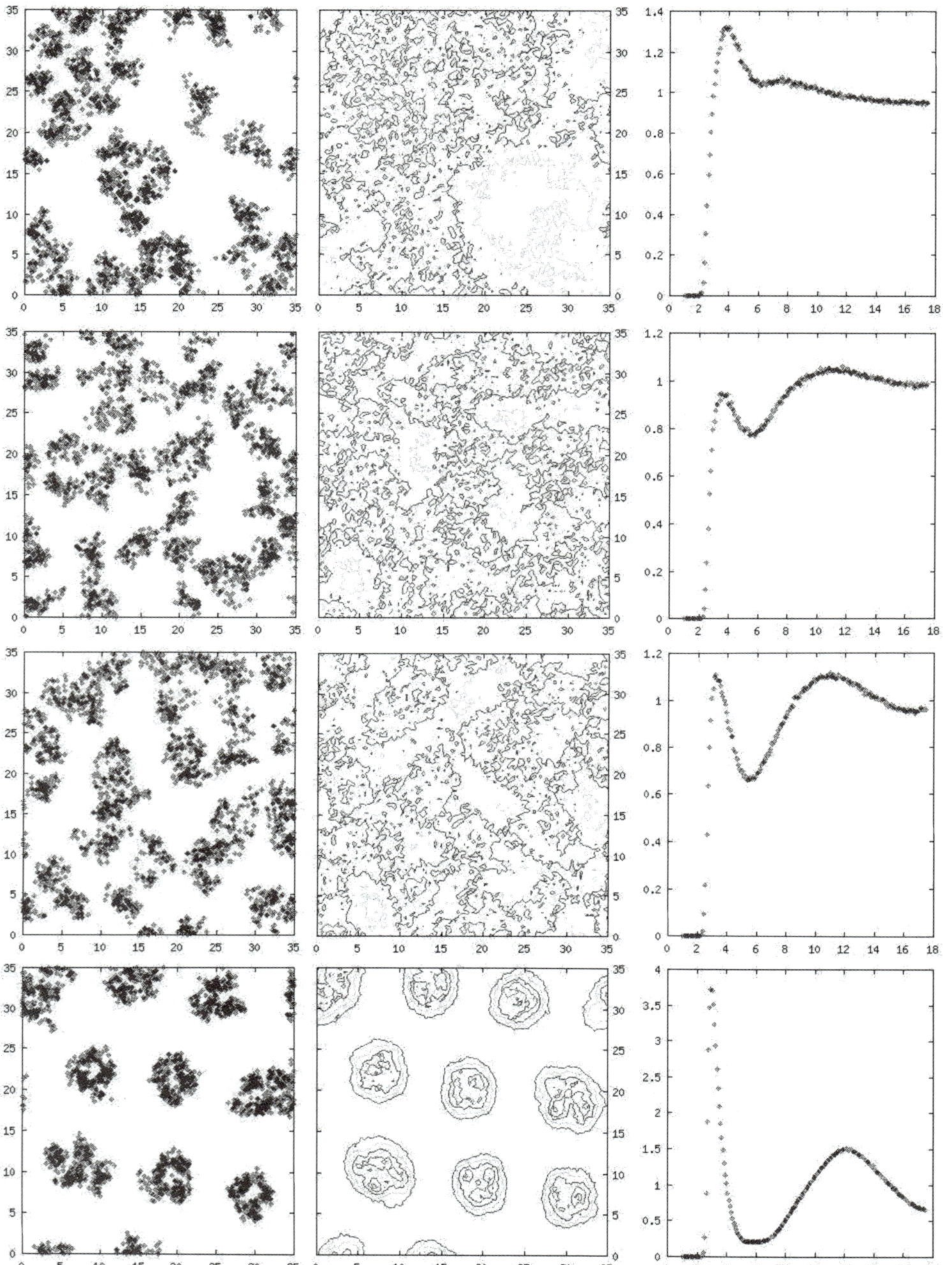

Fig. 4. A thermalized space-time configuration, the contour plot and the pair distribution at surface density $0.0245 Å^{-2}$, $T = 500mK$ and $\alpha = 0$ (top row), $\alpha = 50° KÅ$ (second from the top row), $\alpha = 100° KÅ$ and $\alpha = 300° KÅ$ (bottom row)

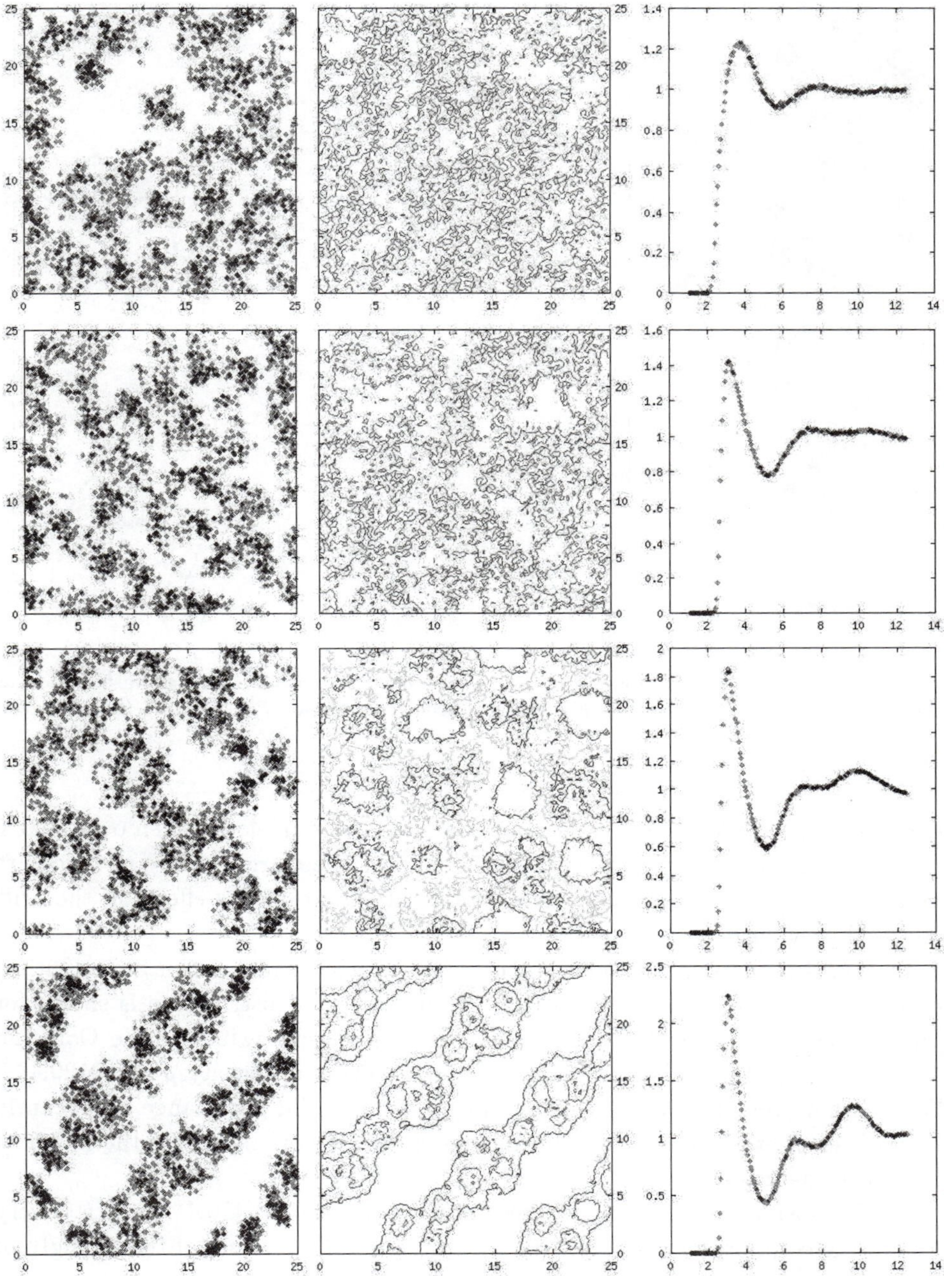

Fig. 5. A thermalized space-time configuration, the contour plot and the pair distribution at surface density $0.0480\AA^{-2}$, $T = 500mK$ and $\alpha = 0$ (top row), $\alpha = 200°K\AA$ (second from the top row), $\alpha = 300°K\AA$ and $\alpha = 350°K\AA$ (bottom row)

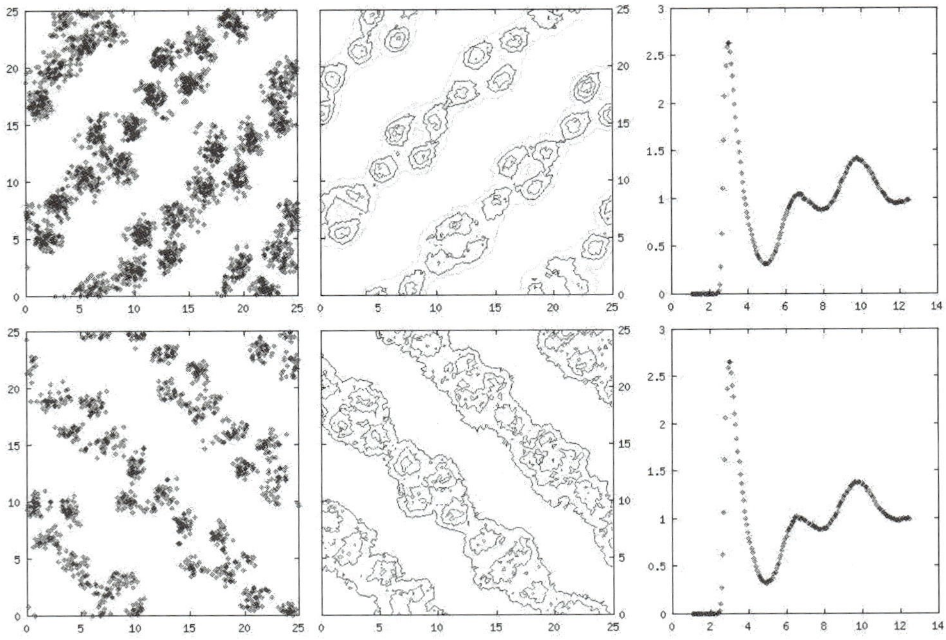

Fig. 6. A thermalized space-time configuration, the contour plot and the pair distribution at surface density $0.0480\AA^{-2}$, $\alpha = 450° K\AA$ and $T = 500mK$ (top row), $T = 1000mK$ (bottom row)

they are those of a Wigner crystal of clusters, the contour plots, taken over 10,000 configurations, indicate that the system is uniform. The larger values of α require averaging over much larger number of configurations to see the effects of the drift discussed previously.

In Fig. 4 a thermalized snapshot of a space-time configuration along with a contour plot of the probability distribution and the pair distribution is shown for a higher density of $0.0245\AA^{-2}$, $T = 500mK$ and various values of the Coulomb repulsion parameter α. The situation is similar to that of density $\rho = 0.0120\AA^{-2}$. We note, however, that while the inter-cluster distance did not change significantly the number of atoms in each cluster has increased by approximately the factor of the density increase.

In Fig. 5 we present the same three functions for $\rho = 0.0480\AA^{-2}$ which is very close to the equilibrium density of the neutral liquid. Low values of the strength of Coulomb repulsion term have small influence on these functions including the pair distribution function. We need to increase to a value $\alpha = 200° K\AA$ in order to see an influence of the repulsion on $g(r)$ comparable to that occurring at $\alpha = 50° K\AA$ for lower values of density. The reason for this insensitivity is the fact at the equilibrium density the compressibility of the fluid is finite, while that of the phase separated state is close to zero.

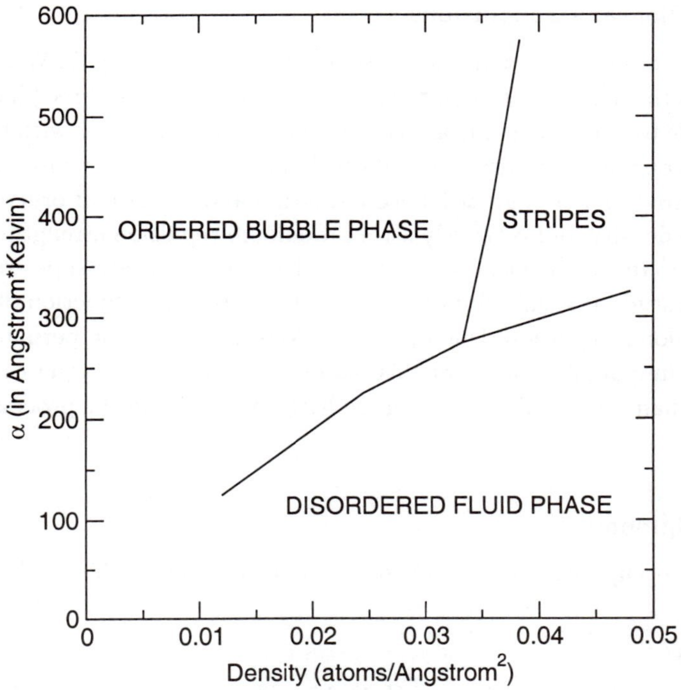

Fig. 7. The phase diagram of a submonolayer Van der Waals-Bose quantum film where an additional long-range Coulomb interaction has been added.

Notice as the values of α increases the clustering that occurs is clusters of vacancies which tend to order on a triangular superlattice. We have not clearly observed an ordered superlattice of clusters of vacancies. This tendency for ordering occurs near the equilibrium density where in a phase separated state there is much more fluid than vacant space. The Coulomb repulsion tends to compress and spread the fluid equally in all directions thus creating clusters of particles or of vacancies or stripes. The stripes occur at densities near the equilibrium density and for large values of α. Notice that $g(r)$ has three peaks when the stripes form or tend to form in Fig. 5. The short-distance peak is that which corresponds to the Van der Waals minimum, the peak at the longest distance corresponds to the inter-stripe distance and the intermediate peak corresponds to the a further ordering within the stripe. Notice in the contour plots of Fig. 5 that the stripes are formed by merging of clusters and the stripe width is modulated along the stripe and the modulation periodicity defines this third peak structure of the pair distribution function.

In Fig. 6 we compare the situation for a larger value of $\alpha = 450°K\AA$ for the equilibrium density and at temperature $T = 500mK$ and $T = 1000mk$. Notice that the detailed ordering of particles with a stripe begins to go away at the higher temperature because of thermal fluctuations. However, the peaks of the pair distribution function remain unchanged.

4. Phase Diagram-Conclusions

Our results are summarized on the phase diagram shown in Fig. 7. Within our large error bars in the determination of the phase boundaries of the three phases, we have not been able to find any significant shift in the phase boundaries with temperature. The phase boundaries separate a uniform fluid phase from a phase of bubbles or clusters ordered in a triangular lattice formation and a phase of ordered stripes at high surface density and relatively strong Coulomb repulsion strength.

In order to draw conclusions relevant to the case of cuprate superconductors we need to also study the superfluid density of the system as a function of the density and the Coulomb repulsion. Due to the non-local nature of the permutation cycles and the winding number operator this requires a significantly larger computational time scales than the calculation presented here and we are in the process of carrying it out.

Acknowledgments

This work was supported by NASA grants NAG-1773 and NAG-2867.

References

1. V. J. Emery, S. A. Kivelson, and H. Q. Lin, *Phys. Rev. Lett.*, **64**, 475 (1990). S. A. Kivelson, E. Fradkin and V. J. Emery, Nature (London) **393**, 550 (1998).
2. C. S. Hellberg and E. Manousakis, Phys. Rev. **B 61**, 11787 (2000). C. S. Hellberg and E. Manousakis, Phys. Rev. Lett. **78**, 4609 (1997).
3. E. W. Carlson, V. J. Emery, S. A. Kiveslon, D. Ograd, "The Physics of Conventional and Unconventional Superconductors" ed. by K. H. Bennemann and J. B. Ketterson (Springer-Verlag).
4. M. Boninsegni and E. Manousakis Phys. Rev. B 47, 11897-11904 (1993)
5. E. Manousakis, Phys. Rev. B 67, 195103 (2003).
6. D. S. Greywall, Phys. Rev. B **47**, 309 (1993).
7. P. A. Crowell and J. D. Reppy, Phys. Rev. Lett. **70**, 3291 (1993).
8. P. A. Crowell and J. D. Reppy, Phys. Rev. B **53**, 2701 (1996).
9. M. Pierce and E. Manousakis, Phys. Rev. Lett. **81**, 156 (1998). M. Pierce and E. Manousakis, Phys. Rev. B **59**, 3802 (1999). M. Pierce and E. Manousakis, Phys. Rev. B **62**, 5228 (2000). M. Pierce and E. Manousakis, Phys. Rev. Lett. **83**, 5314 (1999).
10. W. E. Carlos and M. W. Cole, Surf. Sci. **91**, 339 (1980).
11. R. A. Aziz *et al.*, Mol. Phys. **77**, 321 (1992).
12. D. M. Ceperley, Phys. Rev. **B 18**, 3126 (1978).

CAN INCOMMENSURATION STABILIZE A SUPERFLUID PHASE OF *PARA*-HYDROGEN?

MASSIMO BONINSEGNI

Department of Physics, University of Alberta

Received 1 December 2004

Extensive Path Integral Monte Carlo simulations of two-dimensional *para*-Hydrogen embedded in a crystalline matrix of Alkali atoms, show no evidence of superfluid behavior at low temperature. Rather, the system is observed to form a commensurate (non superfluid) crystal.

Keywords: Superfluidity; Bose condensation; molecular hydrogen; quantum simulations.

1. Introduction

The speculated[1] Superfluid (SF) phase of bulk *para*-Hydrogen (p-H_2) has so far eluded direct experimental confirmation. Unlike Helium, H_2 solidifies at a temperature ($T \sim 14$ K) significantly higher than that (~ 5 K) at which such phenomena as Bose Condensation and superfluidity, are expected to occur. This is due the depth of the attractive well of the potential between two Hydrogen molecules, significantly greater than that between two Helium atoms.

The question of whether p-H_2 would display a SF transition at low enough a temperature, should a sufficiently clever procedure be imagined to suppress crystallization, is clearly of great fundamental interest. A fair amount of experimental and theoretical effort is thus aimed at identifying ways of stabilizing a liquid phase of p-H_2 at low temperature. Confinement, reduction of dimensionality, and incommensuration, are widely regarded as plausible avenues to achieving that goal.

In particular, the suggestion was made by Gordillo and Ceperley[2] (GC) that SF ought to occur in a strictly two-dimensional (2D) p-H_2 fluid embedded in a crystalline matrix of Alkali atoms. Their contention, based on Path Integral Monte Carlo (PIMC) simulations, is that the presence of the underlying lattice of foreign atoms, incommensurate with the equilibrium crystal structure of pure p-H_2, causes a substantial reduction of the equilibrium density of the 2D fluid of p-H_2 molecules, stabilizing a liquid phase, which would then turn superfluid at $T \sim 1$ K.

While undoubtedly intriguing, the above predictions are based on simulations of very small systems (of the order of 10-15 p-H_2 molecules). While it is in principle possible to perform finite-size scaling of the results, attempting extrapolation to the thermodynamic limit, this procedure is not always reliable, particularly with the typical statistical uncertainties that occur in Monte Carlo simulations. Thus, it seemed worthwhile to us to try and assess the robustness of GC's contention by performing simulations of the same system, with at least an order of magnitude more particles. Seven years after GC's original work, such simulations can be performed rather comfortably on commonly available workstations.

2. Calculation

We model our system of interest exactly as in Ref. 2, namely, as a collection of pointlike quantum particles (N p-H_2 molecules) confined to moving in 2D. The accepted Silvera-Goldman[3] potential is used to describe pairwise interactions among p-H_2 molecules. The system also includes a finite concentration of impurities, represented by fixed particles placed at regular lattice positions. The concentration of impurities is set at 0.01155 Å^{-2}, and the interaction between a p-H_2 molecule and an impurity is described by a Lennard-Jones potential with parameters[4] $\epsilon = 9.54$ K and σ=3.75 Å, just like in Ref. 2.

We studied physical properties of this system in the range of temperature 1-4 K, using PIMC simulations based on the standard procedure.[5,6] Our algorithm also includes the all-important sampling of permutations of p-H_2 molecules, necessary to study SF. We utilized three simulation cells in our study, all with periodic boundary conditions; the smallest one is identical with that used by GC, namely 20 Å×17.32 Å, which corresponds to four impurities; the largest is 50 Å×51.96 Å, which corresponds to thirty impurities (in all cases impurities are arranged on a triangular lattice[7]).

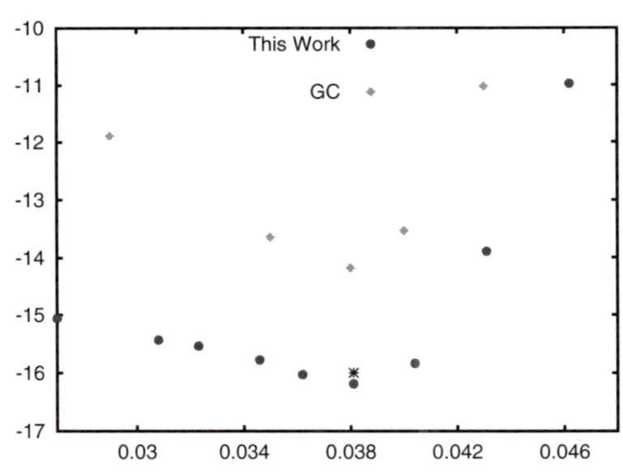

Fig. 1. Energy per p-H_2 molecule (in K) at different 2D densities (in Å^{-2}). The concentration of impurities is 0.01155 Å^{-2}. Circles: this work, T=2 K, 30 impurities. Diamonds: Ref. [2], T=1 K, 4 impurities. Star: This work, T=2 K, 13 p-H_2 molecules. Statistical errors are smaller than the symbols.

Fig. 1 shows computed energetics of 2D p-H_2 films at different 2D densities. Our results are for a temperature T=2 K; we find that the energy estimates are essentially unchanged below $T\sim$ 4 K. Comparison of our results with those of GC shows broad agreement, as far as the general trend and the location of the equilibrium density (~ 0.0381 Å^{-2}) are concerned. There are, however, significant numerical discrepancies between our results and GC's, ours being consistently lower by a few K. This is puzzling, particularly considering that our estimates are obtained at a higher temperature. Although our results pertain to systems of size 7-8 times greater than those of GC's calculation, a numerical discrepancy of this magnitude cannot be attributed to finite-size corrections. In order to illustrate this point, in Fig. 1 we show our energy estimate (star) at the equilibrium density, for a system of 13 p-H_2 molecules, i.e., identical in size with that of GC. Such estimate is in close agreement with the one that we obtain for a larger (99 particles) system.

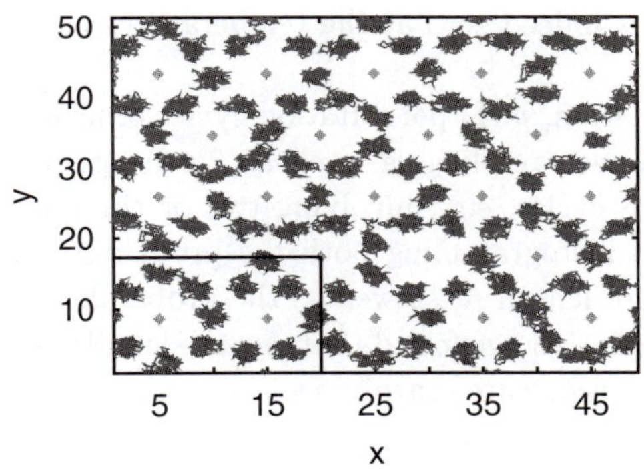

Fig. 2. Instantaneous PIMC configuration of 99 p-H_2 molecules at the equilibrium 2D density. The positions of all molecules at each one of 320 imaginary time slices are shown as discrete paths. Diamonds represent fixed impurities. The smaller rectangle on the lower left part corresponds to the simulation cell used by GC (Ref. [2]).

Our results also suggest a different interpretation than that offered by GC regarding the structure of the equilibrium phase. Fig. 2 shows an instantaneous configuration of 99 p-H_2 molecules at the equilibrium density. The arrangement of p-H_2 molecules on a regular (essentially a *kagome'* lattice) crystal is clear, and can be confirmed by a calculation of the static structure factor for a lattice vector associated to such crystalline arrangement. Little or no overlap is seen among paths, suggesting that permutations are strongly suppressed, which is typical of quantum crystals, and has been consistently found to be the case in PIMC studies of low-dimensional p-H_2 at low temperature, even in low density metastable phases.[6,8,9]

Based on our results, we contend that the equilibrium phase of the system is *not* a liquid, as argued in Ref. [2]; if this system should indeed display superfluidity, it would be a *supersolid*, based on the commonly accepted definition of a supersolid as a phase of matter simultaneously featuring long-range crystalline order and dissipationless flow. The issue to address is whether this system is indeed a superfluid, as proposed in Ref. [2].

Using the "winding number" estimator[5], we have computed the p-H_2 superfluid fraction ρ_s at the equilibrium density for systems of 13, 53 and 99 p-H_2 molecules. The smallest system is enclosed inside the rectangular cell shown in Fig. 2 (lower left corner). For the 13-molecule system, we find a finite value of ρ_s at T=1 K, comparable to that found in Ref. [2]; however, we obtain $\rho_s = 0$ as the system size is increased to 53 and 99 molecules.

In and of itself, the fact that $\rho_s \to 0$ in PIMC simulations of finite systems of increasing size, is not sufficient to conclude that $\rho_s = 0$ in the thermodynamic limit. For, the superfluid signal in PIMC is related to long cycles of permutation of p-H_2 molecules, which are sampled less and less efficiently as the system size is increased. Thus, one may simply be facing an ergodicity problem. However, a careful analysis of the results leads us to surmise that, in this case, the superfluid density in the thermodynamic limit is indeed zero at the temperatures considered here (down to 1 K).

In our PIMC simulation, p-H_2 permutation cycles of arbitrary length can occur as 1) the result of direct sampling, as well as 2) the global result of a number of smaller permutation cycles, e.g., involving two or three particles only (see, for instance, Ref. [5]). On histogramming both the probability of acceptance $P_{acc}(n)$ of permutation cycles of length n, as well as the probability of occurrence $P(n)$ of those cycles in the simulation, we found that, for the smaller system (13 molecules), both probability distributions are sharply peaked at $n = 1$ and $n = 5$. Permutations of two and three particles, instead, occur exceedingly rarely.

The superfluid signal obtained for a system of 13 p-H_2 molecules is clearly related to the cycles of length 5, i.e., the only ones that can result in a nonzero winding. These cycles exclusively occur through direct sampling, rather than as a result of smaller cycles.

Looking at Fig. 2 (small rectangular box in the lower left corner), one can see that permutation cycles involving 5 particles, spanning the entire system, can be obtained, both horizontally as well as vertically, simply by having each particle "hop" into the position of one that is immediately adjacent, which in turn does the same, in the same direction. The last particle in the cycle hops into the position of the first one, across the periodic boundary. By examining configurations generated by the simulations, we convinced ourselves that cycles of length 5 occur exclusively through the simple mechanism described above. In our simulations with greater numbers of molecules, these cycles never occur; with 53 particles, the corresponding cycles would consist of 10 particles, whereas with 99 particles a cycle of the above type would involve 12-15 particles. Although our algorithm can sample such long cycles, acceptance is zero.

There are, in our view, two reasons to believe that these cycles will not contribute a finite superfluid signal at any finite temperature, in the thermodynamic limit. Firstly, because they only occur as a result of direct sampling, rather than through many, more frequently sampled smaller cycles; secondly, because they are essentially one-dimensional in character.

3. Conclusions

Based on our study of two-dimensional *para*-Hydrogen embedded in a crystalline matrix of Alkali atoms, modeled in exactly the same way as in a previous study, we conclude that this system is *not* a candidate for observing superfluidity in p-H_2. At low temperature, the system forms instead a 2D crystal, commensurate with the underlying lattice of impurities. We have presented results for triangular impurity lattices, but the same results were seen with other lattices as well (e.g., rectangular). The finite superfluid signal obtained by other authors can be attributed, in our view, to the finite size of the simulation cell used.

Acknowledgments

This work was supported in part by the Petroleum Research Fund of the American Chemical Society under research grant 36658-AC5, and by the Natural Science and Engineering Research Council of Canada under research grant G121210893.

References

1. V. L. Ginzburg and A. A. Sobyanin, JETP Letters **15**, 242 (1972).
2. M. C. Gordillo and D. M. Ceperley, Phys. Rev. Lett. **79**, 3010 (1997).
3. I. F. Silvera and V. V. Goldman, J. Chem. Phys. **69**, 4209 (1978).
4. It has been suggested (C. Cazorla and J. Boronat, J. Low Temp. Phys. **134**, 43 (2004)) that these settings may not apply to any known interaction or substrate. However, we regard it as an interesting problem to identify conditions (if any) under which SF is possible, even though they may be somewhat "artificial".
5. D. M. Ceperley, Rev. Mod. Phys. **67**, 279 (1995).
6. M. Boninsegni, Phys. Rev. B **70**, 125405 (2004).
7. Some of the calculations performed in Ref. 2 had impurities arranged on a rectangular lattice, but the results on rectangular and triangular lattices are found to be qualitatively similar.
8. K. Nho and E. Manousakis, Phys. Rev. B **65**, 115409 (2002), and *ibid.* **67**, 195411 (2003).
9. M. Boninsegni, Phys. Rev. B **70**, in press; cond-mat 0409301 (2004).

ANALYSIS OF THE INTERATOMIC POTENTIAL OF THE HELIUM SYSTEMS

SEBASTIAN UJEVIC* and S. A. VITIELLO†

Instituto de Física, Universidade Estadual de Campinas,
13083 Campinas–São Paulo, Brazil
**sut@ifi.unicamp.br*
†vitiello@ifi.unicamp.br

Received 1 December 2004

The effects of interatomic potentials in the equation of state of ^{4}He and other properties of the system are investigated. A multi-weight diffusion Monte Carlo method is applied in order to compute very small energies differences with great accuracy. From our analysis we identify the best current description for the helium systems.

Keywords: Liquid helium; solid helium; interatomic potential; equation of state.

1. Introduction

In the investigation of the helium systems a considerable effort continues to be placed on the understand of its interatomic potential.[1–3] In this work we extend our analysis[4,5] of this potential by considering various *ab initio* two-body potentials for the description of the He-He interactions and a three-body potential that is needed to correctly describe the system in the liquid and solid phases. The level of accuracy reached by the *ab initio* potentials makes interesting to have a quantitative comparison of them as a function of the system density. It is also our aim to discuss in a quantitative way how some of the properties of interest of the helium systems change with the inclusion of each one of the terms of a three-body interaction that considers triple-dipole and exchange effects.

In the description of bulk helium we have considered three *ab initio* two-body potentials. The first one was proposed by Aziz[1] (V_{HF}) and coworkers. This potential has an overall damped Hartree-Fock-dispersion form. The second was proposed by Korona[2] (V_p) and coworkers, where infinite order symmetry adapted perturbation theory is applied to model the intermediate atomic interactions. Its analytic form uses the damping function proposed by Tang-Toennies.[6] The last potential considered is only slightly different from the V_p potential. It is included a retardation function[3] in the dipole-dipole dispersion term and as a consequence the potential width is reduced. It will be denoted as V_{pr}.

It is known that three-body effects can not be neglected in quantitative investigations of many body systems.[7] In this work we have consider the three-body potential of Cohen and Murrell.[8] It has two terms, the triple-dipole term of Axilrod–Teller–Muto[9,10] (V_D) and an exchange term (V_J). Hereafter the subscript 3 in any of the above two-body potential denotes the inclusion of both terms, *i.e.*, $V_{i3} = V_i + V_D + V_J$.

In this work we have used the multi-weight DMC method (MWDMC) to compute how the potentials we have considered modify the equation of state of ^{4}He in the liquid and solid phases. This is an extension we have proposed[4] to the standard DMC method. It allows the calculation of small energy differences with great accuracy. At each density only a single set of configurations is used to evaluate the properties associated to all the potentials and thus the quantities of interest are estimated in a correlated fashion. A detailed account of the MWDMC has been published elsewhere.[4] In the next section we give details of the simulations and present our results. The last section is devoted to our conclusions.

2. The Simulations and Results

We studied bulk helium in the liquid and solid phases by imposing periodic boundary conditions on a system of 108 atoms. The solid phase was studied using a *fcc* structure in a cubic cell. Tail corrections were performed for the two-body potentials; no tail corrections were considered for the high order terms of the potentials.

In the liquid phase a guiding function of the Jastrow form was used: $\Psi_J(R) = \prod_{i<j} f(r_{ij})$, where $f(r_{ij}) = \exp[-u(r_{ij})/2]$. It correlates pairs of particles with a pseudo-potential of the McMillan form, $u(r_{ij}) = (b/r_{ij})^5$, where b is a parameter. For the solid phase a Nosanov-Jastrow form function $\Psi_{NJ}(R) = \Psi_J(R)\Phi(R)$ was considered, $\Phi(R) = \prod_i \exp\left[-\frac{C}{2}(\mathbf{r}_i - \mathbf{l}_i)^2\right]$ represents a mean field term that explicitly localize the particles around given lattice sites $\mathbf{l}_i$.

We have constructed equation of state (EOS) for the liquid and solid phases for all the potentials considered in this work. In Table 1 typical energies obtained in the calculations are compared with experimental values. We have also fitted analytical equations of state for the liquid and solid phases using a function of the form $E(\rho) = E_0 + A\left[(\rho-\rho_0)/\rho_0\right]^2 + B\left[(\rho-\rho_0)/\rho_0\right]^3$, where E_0, ρ_0, A and B are the fitting parameters. The results for the liquid phase are plotted in Fig. 1. Contributions of the triple-dipole (E^D) and exchange (E^J) terms to the total energy when the V_{HF} and V_{pr} potentials were employed in the calculations are presented in Table 2. For comparison we show also extrapolated results of perturbative calculations performed using configurations generated with the V_{HF} and V_{pr} potentials. The two methods give results that do not always agree within the statistical uncertainty.

The melting and freezing densities results were obtained by a Maxwell double-tangent construction using the analytical equations of state for the liquid and solid phases. In Table 3 we present the melting, freezing and equilibrium densities for all the potentials we have considered. Note how these properties evolves when each terms of the three-body potential are included into the V_{HF} two-body potential. Since we use correlated sampling we can trust the relative values.

Table 1. Total energies per atom in units of K obtained with the V_{p3}, V_{pr3}, V_{HF3} and V_{pr} potentials at the given densities. In the last column we show experimental data or values obtained through an analytical EOS fitted to the experimental data.

ρ (nm^{-3})	V_{p3}	V_{pr3}	V_{HF3}	V_{pr}	Exp.
			Liquid		
19.64	-7.080 ± 0.008	-7.060 ± 0.008	-7.017 ± 0.008	-7.172 ± 0.008	-7.044[a]
21.86	-7.196 ± 0.008	-7.174 ± 0.008	-7.120 ± 0.008	-7.319 ± 0.008	-7.170[b]
24.01	-7.053 ± 0.009	-7.029 ± 0.009	-6.956 ± 0.009	-7.217 ± 0.009	-7.027[b]
26.23	-6.558 ± 0.011	-6.532 ± 0.011	-6.437 ± 0.011	-6.767 ± 0.011	-6.556[a]
			Solid		
29.34	-5.754 ± 0.009	-5.725 ± 0.009	-5.625 ± 0.009	——	-5.789[c]
32.88	-4.263 ± 0.009	-4.231 ± 0.009	-4.085 ± 0.009	-4.644 ± 0.009	-4.317[c]
33.54	-3.863 ± 0.009	-3.830 ± 0.009	-3.675 ± 0.009	——	-3.919[c]
35.27	-2.571 ± 0.010	-2.537 ± 0.010	-2.352 ± 0.010	-3.032 ± 0.010	-2.681[c]

[a]Extrap. value from Ref. 11, [b]Interp. value from Ref. 11, [c]Interp. value from Ref. 12.

Table 2. Energies per particle in units of K associated with the triple-dipole term (E^D) and the exchange term (E^J) at the given densities and potentials. The results are obtained by MWDMC and by extrapolation of perturbative calculation. In the liquid phase the result for the V_{HF} potential at $\rho = 26.23$ nm^{-3} was obtained with 64 bodies.

Two-body		E^D		E^J	
potential	ρ (nm^{-3})	MWDMC	Extr.	MWDMC	Extr.
			Liquid		
V_{HF}	21.86	0.137 ± 0.002	0.1340 ± 0.0004	0.0059 ± 0.0003	0.0070 ± 0.0001
	26.23	0.216 ± 0.001	0.2128 ± 0.0002	0.0070 ± 0.0004	0.0080 ± 0.0001
V_{pr}	21.86	0.139 ± 0.001	0.1370 ± 0.0002	0.0061 ± 0.0004	0.0073 ± 0.0001
	26.23	0.227 ± 0.001	0.2249 ± 0.0002	0.0083 ± 0.0003	0.0091 ± 0.0002
			Solid		
V_{HF}	32.88	0.417 ± 0.001	0.4129 ± 0.0002	-0.0052 ± 0.0003	-0.0029 ± 0.0002
	35.27	0.509 ± 0.001	0.5035 ± 0.0002	-0.0130 ± 0.0005	-0.0100 ± 0.0003
V_{pr}	32.88	0.418 ± 0.001	0.4138 ± 0.0002	-0.0052 ± 0.0002	-0.0026 ± 0.0002
	35.27	0.508 ± 0.001	0.5018 ± 0.0002	-0.0129 ± 0.0004	-0.0090 ± 0.0003

Table 3. Melting (ρ_m), freezing (ρ_f) and equilibrium (ρ_0) densities for the given potentials. In the last line we present experimental values.

Potential	ρ_f (nm^{-3})	ρ_m (nm^{-3})	ρ_0 (nm^{-3})
V_{p3}	25.98 ± 0.18	28.98 ± 0.18	21.820 ± 0.019
V_{pr3}	25.98 ± 0.18	28.98 ± 0.18	21.804 ± 0.019
V_{HF3}	25.90 ± 0.21	28.91 ± 0.22	21.681 ± 0.017
$V_{HF} + V_D$	25.92 ± 0.21	28.94 ± 0.22	21.692 ± 0.017
V_{HF}	25.83 ± 0.21	28.95 ± 0.22	21.999 ± 0.017
Exp.	25.970 ± 0.005[13]	28.677[14]	21.834[13]

3. Conclusions

In order to determined the best interatomic potential to be used in simulational studies of helium atoms systems, we will now discuss how our results compare

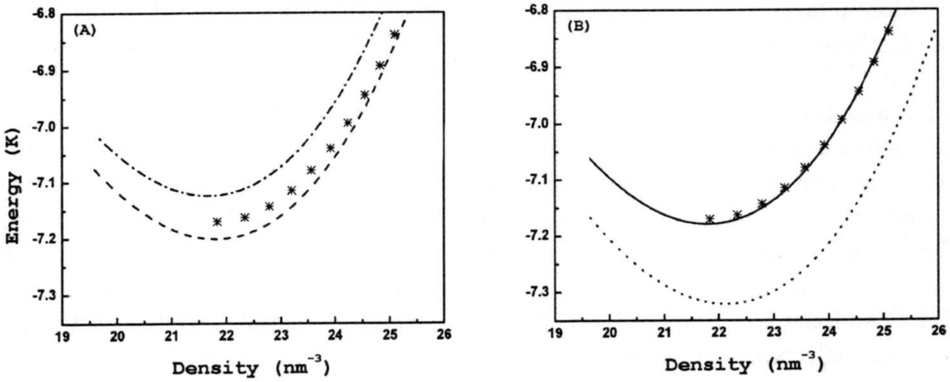

Fig. 1. (A) Analytical equations of state for the liquid phase obtained using the V_{p3} (dashed line) and V_{HF3} (dot-dashed line) potentials. (B) Comparison of the analytical EOS determined using the V_{pr} potential (dotted line) with this potential plus the three-body terms, V_{pr3} (solid line). In both figures the asterisks stands for the experimental data of Ref. 11.

with the experimental values. We see from Fig. 1 (Table 1) that a pure two-body interatomic potential would bind too much the system. In the liquid phase the results obtained using the V_{pr3} potential are in a remarkable agreement with the experimental values. Also in this phase we can see that the V_{p3} potential binds too much the system at low densities. For the solid phase none of the considered potentials give results that are in agreement with the experimental values. The two-body potentials together with the Cohen-Murrell potential are not attractive enough to bind the system as much as needed.

The results of the equilibrium density (Table 3) are in agreement with the experimental value only for the V_{p3} and V_{pr3} potentials. All potentials give results for the melting and freezing densities, within their statistical uncertainties, in agreement with the experimental values. Thus to discriminate these potentials using these quantities we need more precise results.

Regarding the three-body potential, its inclusion lowers the value of the equilibrium density in calculations where the two-body potential is assumed to be V_{HF}. This behavior of the result is well came because the V_{HF} potential used without a three-body component give a equilibrium density too high. However the computed value of this quantity becomes too small. With respect to the melting and freezing density the inclusion of the triple-dipole term in the V_{HF} potential displace the computed values of these quantities in the direction of the experimental value. However as in the case of the equilibrium density, the exchange term does not improve the freezing density results. It improves by very little the melting density. We stress that we are dealing with correlated results that allows us to trust the small differences we are observing. Moreover if we look at the contribution of the exchange term to the total energy of the system (Tables 1 and 2), we see that only at the solid phase it slightly starts to improve the results.

More information about the potentials can be eventually obtained by analyzing the isothermal compressibility $K(\rho) = (1/\rho)(\partial\rho/\partial P)_T$, the sound velocity $c(\rho) = (mK\rho)^{-1/2}$ and the pressure $P(\rho) = \rho^2 (\partial E/\partial\rho)$. The computed values for these quantities favors the use of the V_{p3} and V_{pr3} potentials. However these results do not help us to differentiate the two potentials.[5] The computed quantities are too closed to each other and to the experimental value.

By taking into account the above discussion, we believed that the best current choice for the description of the systems of helium atoms is accomplished using the interatomic potential V_{prD}. It is formed by the *ab initio* two-body potential of Ref. 3 and the triple-dipole term of the three-body potential of Ref. 8. Although exchange effects can be of importance in the crystallization process, it is possible that the term with which we have experimented can not describe these effects at the accuracy that our calculations can probe. Although it improves few results, others are worsened by its inclusion, see for instance the freezing and equilibrium densities in Table 3.

If we consider the high accuracy of the available *ab initio* two-body potentials, our results suggest that for a better understanding of the helium systems, further work should be concentrated in the three-body part of its interatomic potential.

Acknowledgments

This work was conducted, in part, using the facilities of the "Centro Nacional de Processamento de Alto Desempenho em São Paulo". SU thanks a scholarship from "Fundação de Amparo à Pesquisa do Estado de São Paulo - FAPESP".

References

1. R. A. Aziz, A. R. Janzen, and M. R. Moldover, *Phys. Rev. Lett.* **74**, 1586 (1995).
2. T. Korona, H. L. Williams, R. Bukowski, B. Jeziorski, and K. Szalewicz, *J. Chem. Phys.* **106**, 5109 (1997).
3. A. R. Janzen and R. A. Aziz, *J. Chem. Phys.* **107**, 914 (1997).
4. S. Ujevic and S. A. Vitiello, *J. Chem. Phys.* **119**, 8482 (2003).
5. S. Ujevic and S. A. Vitiello, *unpublished.*
6. K. T. Tang and J. P. Toennies, *J. Chem. Phys.* **80**, 3726 (1984).
7. M. J. Elrod and R. J. Daykally, *Chem. Rev.* **94**, 1975 (1994).
8. M. J. Cohen and J. N. Murrell, *Chem. Phys. Lett.* **260**, 371 (1996).
9. B. M. Axilrod and E. Teller, *J. Chem. Phys.* **11**, 299 (1943).
10. Y. Muto, *Proc. Phys. Math. Soc. Jpn.* **17**, 629 (1943).
11. R. de Bruyn Ouboter and C. N. Yang, *Physica.* **B144**, 127 (1987).
12. D. O. Edwards and R. C. Pandorf, *Phys. Rev.* **140A**, 816 (1965).
13. J. E. Berthold, H. N. Hanson, H. J. Maris, and G. M. Seidel, *Phys. Rev.* **B14**, 1902 (1976).
14. A. Driessen, E. van der Poll, and I. F. Silvera, *Phys. Rev.* **B33**, 3269 (1986).

Nuclear Physics and QCD

QUANTUM PHASE TRANSITIONS IN MESOSCOPIC SYSTEMS

F. IACHELLO

Center for Theoretical Physics, Sloane Physics Laboratory,
Yale University, New Haven, CT 06520-8120, USA
francesco.iachello@yale.edu

Received 1 December 2004

Shape phase transitions in nuclei, an example of quantum phase transitions in mesoscopic systems, are briefly reviewed. Spectral signatures of critical points in finite quantal systems are discussed and experimental examples are shown.

Keywords: Quantum phase transitions; interacting Boson model; mesoscopic systems.

1. Quantum Phase Transitions

A Quantum Phase Transition (QPT) is a phase transition that occurs as a function of a coupling constant, g, (called control parameter), that appears in the quantum Hamiltonian, H, that describes the system

$$H = \alpha_1 H_1 + \alpha_2 H_2 = c(H_1 + gH_2), \tag{1}$$

where H_1 and H_2 describe two different phases of the system, and c is a scale factor. Associated with a phase transition there is an *order parameter*, the expectation value of a suitably chosen operator that describes the state of the system $\langle P \rangle$. An example from condensed matter physics is the Ising model in a transverse magnetic field,[1] which has a second order QPT at $g = g_c = \frac{1}{2}$. The order parameter is the magnetization $M \propto \langle \hat{\sigma}^z \rangle$ and the two phases are: (i) ferromagnet for $g < g_c$ and (ii) paramagnet for $g > g_c$.

2. Quantum Phase Transitions in Finite Systems

Many finite systems (molecules, nuclei, atomic clusters, finite polymers, ...) are characterized in their equilibrium configuration by a shape. These shapes are in many case rigid. However, there are several situations in which the system is rather floppy and undergoes a phase transition between two different shapes. Shape phase transitions are examples of QPT. Challenging problems are: (i) how to study shape phase transitions in finite systems and (ii) how to describe their properties in the phase transition region and in particular at the phase transition point.

A method that has been found very useful to decribe quantum phase transitions in finite systems is that of modeling the many-body system on a compact algebraic space in n dimensions, $U(n)$, where n is the number of degrees of freedom of the individual particles. States are then characterized by the irreducible representations of $U(n)$, labelled by the total number of particles N. The representations are totally symmetric for bosons and totally antisymmetric for fermions. Many-body systems composed of a mixture of bosons and fermions have also been considered. The modeling for these systems is done on a compact algebraic superspace in n fermionic and m bosonic dimensions, $U(n/m)$. States are characterized by the totally supersymmetric representations of $U(n/m)$, labelled by the total number of bosons plus fermions, $N = N_B + N_F$.

2.1. *Quantum phase transitions in nuclei*

An algebraic model of nuclei is provided by the Interacting Boson Model.[2] This is a description of nuclei in terms of correlated pairs of nucleons with angular momentum $J = 0$ and $J = 2$ (s- and d-wave pairing), treated as bosons, and denoted by s and $d_\mu (\mu = 0, \pm 1 \pm 2)$. Since there are six components, the algebraic structure of this model is $U(6)$. States are labelled by the totally symmetric representations $[N]$ of $U(6)$. N is the number of bosons, taken to be the number of valence particle or hole pairs.

The first step in the study of QPT is the construction of the phase diagram. This is done by identifying all possible dynamic symmetries of the system. In the Interacting Boson Model there are three dynamic symmetries, corresponding to the breaking of $U(6)$ into its subalgebra chains,

$$\begin{array}{ll} U(6) \supset U(5) \supset SO(5) \supset SO(3) \supset SO(2) & \text{(I)} \\ U(6) \supset SU(3) \supset SO(3) \supset SO(2) & \text{(II)} \\ U(6) \supset SO(6) \supset SO(5) \supset SO(3) \supset SO(2) & \text{(III)}. \end{array} \tag{2}$$

The next step is to write down a Hamiltonian that is a combination of Hamiltonians describing the dynamic symmetries. For the Interacting Boson Model a suitable Hamiltonian is[3]

$$H = \varepsilon_0 [(1 - \xi) \hat{n}_d - \frac{\xi}{4N} \hat{Q}^\chi \cdot \hat{Q}^\chi] \tag{3}$$

where $\hat{n}_d = (d^\dagger \cdot \tilde{d})$; $\hat{Q}^\chi = (d^\dagger \times \tilde{s} + s^\dagger \times \tilde{d})^{(2)} + \chi (d^\dagger \times \tilde{d})^{(2)}$.There are here two control parameters ξ, χ. When $\xi = 0$, the system has symmetry $U(5)$. When $\xi = 1, \chi = -\frac{\sqrt{7}}{2}$, the symmetry is $SU(3)$, and when $\xi = 1, \chi = 0$, the symmetry is $SO(6)$.

The final step is to associate shapes to the algebraic structure by introducing coherent states $\mid N; \alpha_\mu \rangle = (s^\dagger + \sum_\mu \alpha_\mu d_\mu^\dagger)^N \mid 0 \rangle$.[4,5] The classical variables α_μ live on the coset space $U(6)/U(5) \otimes U(1)$. By making use of rotational invariance, one can eliminate the Euler angles and rewrite the coherent states in term of (intrinsic) Bohr variables β and γ

$$(s^\dagger + \beta [\cos \gamma d_0^\dagger + \frac{1}{\sqrt{2}} \sin \gamma (d_{+2}^\dagger + d_{-2}^\dagger)])^N \mid 0 \rangle. \tag{4}$$

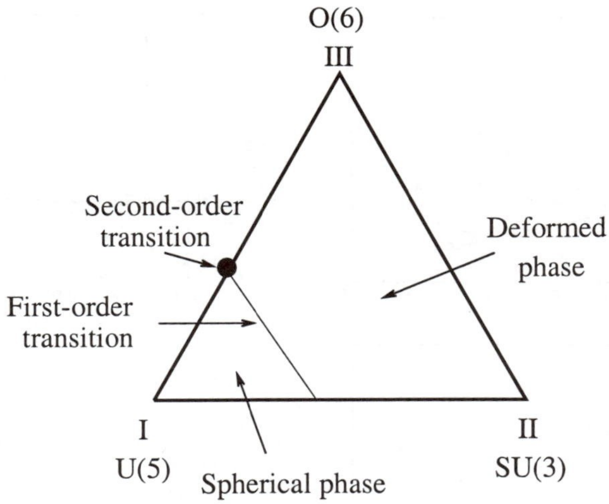

Fig. 1. Phase diagram of IBM-1.

Shapes of nuclei (seen as liquid drops) can be visualized by introducing a surface $R = R_0(1 + \sum_\mu \alpha_\mu Y_{2\mu}(\theta, \varphi))$. The surface is an ellipsoid.

The ground state energy functional (often called the potential) is

$$E(\beta, \gamma) = \frac{\langle N; \beta, \gamma \mid H \mid N; \beta, \gamma\rangle}{\langle N; \beta, \gamma \mid N; \beta, \gamma\rangle}. \tag{5}$$

By minimizing E with respect to β and γ one obtains the equilibrium configuration. For the three dynamic symmetries, one obtains: (I) $U(5)$: Spherical shape; (II) $SU(3)$: Axially deformed shape; (III) $SO(6)$: γ-unstable shape. The equilibrium values β_e and γ_e are the order parameters.

Shape phase transitions can be studied by means of an algorithm introduced by Gilmore.[6] By considering Hamiltonians that are combinations of invariant (Casimir) operators of two chains and minimizing the energy functional $E(N; \beta, \gamma)$ as a function of β, γ, one can determine $E_{\min}$. By studying the behavior of $E_{\min}$ and its derivatives as a function of the coupling constants (control parameters, ξ, χ), one can determine the order of the phase transition (Erhenfest classification). When $N \to \infty$ discontinuities appear at the critical values of the order parameters. When the discontinuity is in $E_{\min}$, the transition is called of $0th - order$, when it is in $\frac{\partial E_{\min}}{\partial \xi}$ is of $1st - order$, when in $\frac{\partial^2 E_{\min}}{\partial \xi^2}$ of $2nd - order$, etc. Gilmore's algorithm was used in the early 1980's to study the phase transitions of the Interacting Boson Model.[5,7] The results was: A: $U(5) - SO(6)$ (spherical to γ-unstable deformed) 2nd order transition; B: $U(5) - SU(3)$ (spherical to axially deformed) 1st order transition; C: $SO(6) - SU(3)$ (γ-unstable deformed to axially deformed) no phase transition (crossover). The corresponding phase diagram of nuclei in IBM is shown in Fig. 1.[7]

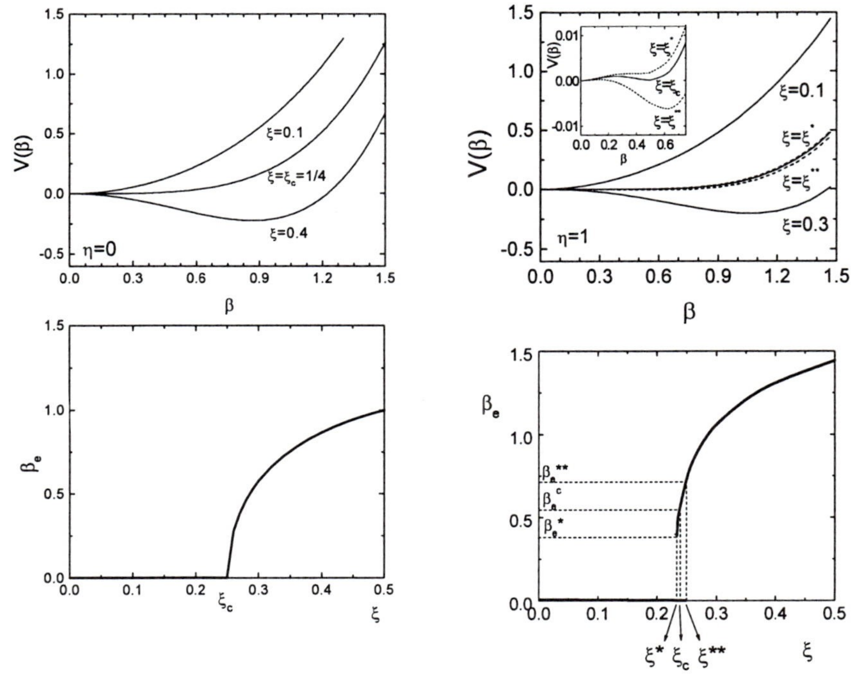

Fig. 2. Behavior of the potential (top panel) and the classical order parameter, β_e, (bottom panel) around the critical point. Left: the $U(5) - SO(6)$ transition; right: the $U(5) - SU(3)$ transition. Adapted from [8].

2.2. *Behavior of the order parameter around the critical point*

2.2.1. *U(5)-SO(6) transition*

This transition (spherical to γ-unstable shapes) is second order. It is characterized by the critical value ξ_c and critical exponent μ. The order parameter is $\beta_e = 0$ for $\xi < \xi_c$ and $\beta_e \propto (\xi - \xi_c)^\mu$ for $\xi \geq \xi_c$. The mean field values are $\xi_c = 0.5$, $\mu = 0.5$, Fig. 2 (left).

2.2.2. *U(5)-SU(3) transition*

This transition (spherical to axially deformed shapes is first order). For first order transitions, there is coexistence of two minima. There are here three relevant values, the spinodal value, ξ^*, the critical value, ξ_c, and the antispinodal value, ξ^{**}.[8] One can define a spinodal exponent, μ^*. The order parameter behaves as $\beta_e \propto (\xi - \xi^*)^{\mu^*}$ for $\xi \geq \xi^*$. The mean field value is $\mu^* = 0.5$, Fig. 2 (right).

A crucial question for mesoscopic systems is how to define phase transitions in finite N systems. Algebraic methods offer an opportunity to study finite N effects by numerical diagonalization of the Hamiltonian.[8] For second order transitions, one

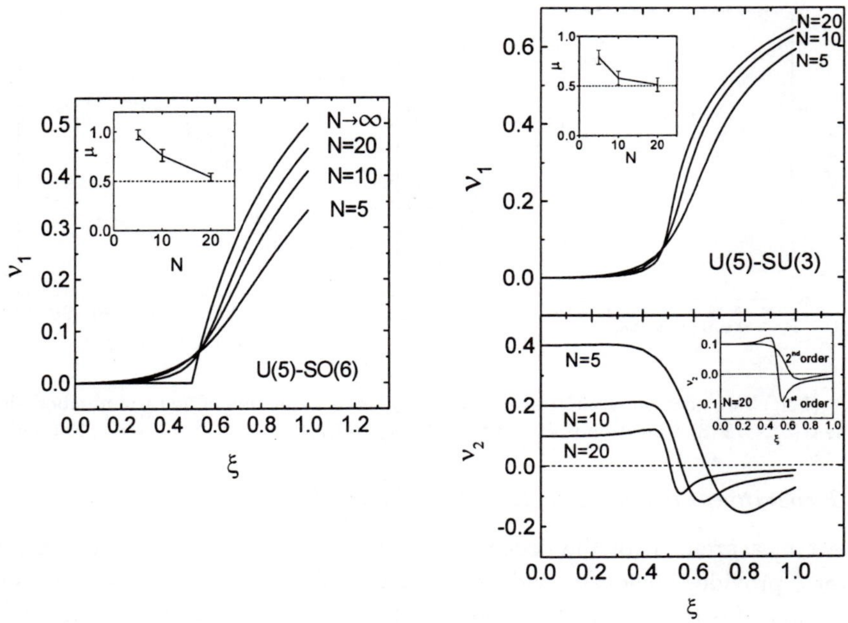

Fig. 3. The quantum order parameters as a function of the control parameter. Left: the $U(5) - SO(6)$ transition; right: the $U(5) - SU(3)$ transition. In the inserts, the critical and spinodal exponents. Adapted from [8].

defines a quantum order parameter

$$\nu_1 = \frac{\langle n_d \rangle_{0_1}}{N} \propto \beta_e^2 \tag{6}$$

It appears, Fig. 3 (left), that for this problem, the main features of QPT persists even for $N \sim 10$. From studies in other systems, described by the algebra $U(n)$, it appears that phase transitional behavior sets in at $nN \sim 50$. A numerical extraction of the critical exponent, gives $\mu = 0.53 \pm 0.03$.

For first order transitions the situation is more complex. Two order parameters are needed. One can define them as

$$\nu_1 = \frac{\langle n_d \rangle_{0_1}}{N}; \quad \nu_2 = \frac{\langle n_d \rangle_{0_2} - \langle n_d \rangle_{0_1}}{N}. \tag{7}$$

Also here, phase transitional behavior sets in at $N \sim 10$, Fig. 3 (right). However, finite N effects smear the phase transition region and make it look as a second order transition. The order parameter ν_2 is barely sufficient to distinguish first from second order transitions. One can extract the spinodal exponent $\mu^* = 0.52 \pm 0.07$. Finite N effects can also be studied analytically.[9]

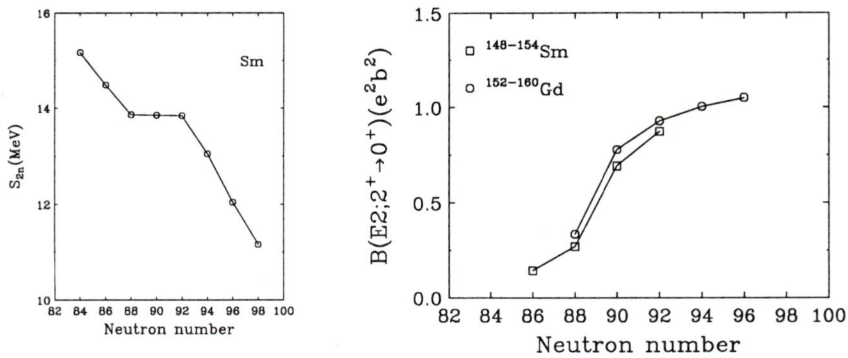

Fig. 4. (Left) Two neutron separation energies in $_{62}Sm$ as a function of neutron number, showing that the $U(5)-SU(3)$ transition is first order. (Right) $B(E2;2_1^+ \to 0_1^+)$ values in $_{62}Sm$ and $_{64}Gd$.

2.3. *Experimental evidence for QPT in nuclei*

A direct measurement of the order parameter β as a function of the control parameter ξ provides obviously evidence for QPT. This direct measurement is often not possible. Instead, indirect measurements of a functional of the order parameter as a function of the control parameter are possible. In the Ising model, one can measure the magnetic susceptibility. In nuclei, since ξ and χ are a function of proton and neutron number (or the number of bosons N) one can obtain evidence for QPT by measuring nuclear properties as a function of N. A considerable amount of experimental evidence for the $U(5) - SU(3)$ transition has been accumulated in the Nd-Sm-Gd isotopes, by measuring: (i) Intensities of electric quadrupole transitions $B(E2;2_1^+ \to 0_1^+)$; (ii) Isomer shift $\delta\langle r^2\rangle = \langle r^2\rangle_{2_1^+} - \langle r^2\rangle_{0_1^+}$; (iii) Two-neutron separation energies $S_2(N) = E_B(N+1) - E_B(N)$; (iv) Isotope shift $\Delta\langle r^2\rangle = \langle r^2\rangle^{(N+1)} - \langle r^2\rangle^{(N)}$. Two of these properties, $B(E2;2_1^+ \to 0_1^+)$ and $S_2(N)$ are shown in Fig. 4. The intensity of $E2$ transitions is particularly relevant, since it is proportional to the order parameter ν_1.

3. Spectral Signatures of Critical Points in Finite Quantal Systems

An important question is what are the properties of systems at the critical point of a quantum phase transition. This question has been addressed recently.[10,11] The main idea is that at the critical point of a phase transition the potential has special properties. For example, at the critical point ξ_c of a second order phase transition the potential is flat and can be replaced by a square well. By solving the differential equation associated with the algebraic structure $U(n)$, one can find the properties of the system at the critical point. They are given in terms of special functions, here Bessel functions. A suitable equation associated with $U(6)$ is the Bohr Hamiltonian[12]

$$H = -\frac{\hbar^2}{2B}\{\frac{1}{\beta^4}\frac{\partial}{\partial\beta}\beta^4\frac{\partial}{\partial\beta} + \frac{1}{\beta^2\sin 3\gamma}\frac{\partial}{\partial\gamma}\sin 3\gamma\frac{\partial}{\partial\gamma}$$

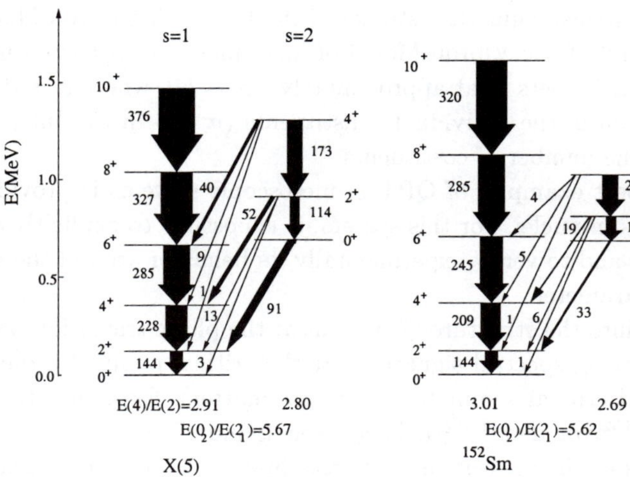

Fig. 5. Comparison between the experimental spectrum of ^{152}Sm and the spectrum of the $X(5)$ solution. Adapted from [14].

$$-\frac{1}{4\beta^2}\sum_{\kappa}\frac{Q_\kappa^2}{\sin^2(\gamma-\frac{2}{3}\pi\kappa)}\}+V(\beta,\gamma). \tag{8}$$

Two solutions of this equation have been worked out:

(I) Critical point of the spherical to γ unstable transition, $U(5)-SO(6)$, called $E(5)$.

The energy eigenvalues are given by

$$E^{(I_c)}(s,\tau,\nu_\Delta,L,M)=E_0+A(x_{s,\tau})^2$$
$$x_{s,\tau}=\text{s-th zero of } J_\nu(z);\quad \nu=\tau+3/2. \tag{9}$$

(II) Critical point of spherical to axially deformed transition, $U(5)-SU(3)$, called $X(5)$.

The energy eigenvalues are given by

$$E^{(II_c)}(s,L,s',K,M)=E_0+A\,(x_{s,L})^2+B(x_{s',K})^2$$
$$x_{s,L}=\text{s-th zero of } J_\nu(z);\quad \nu=\left(\frac{L(L+1)}{3}+\frac{9}{4}\right)^{1/2}$$
$$x_{s',K}=\text{s'-th zero of } J_{\nu'}(z);\quad \nu'=\frac{K}{2}. \tag{10}$$

Examples of nuclei at the critical point of both phase transitions have been found.[13,14] One of them is shown in Fig. 5.

4. Conclusions

It appears from direct numerical diagonalization of the quantum Hamiltonian for finite particle number N that the main features of quantum phase transitions persist in mesoscopic systems. Numerical studies of the Interacting Boson Model of nuclei, reported here, and of the Vibron Model of molecules, not discussed here, indicate that critical behavior sets in at approximately $nN \sim 50$, where n is the number of degrees of freedom of the individual constituents ($n = 6$ in the Interacting Boson Model) and N the number of costituents.

One of the best examples of QPT in mesoscopic systems is provided by shape phase transitions in nuclei. For this system, it is possible to explicitly calculate the order parameter and to verify experimentally its behavior around the critical value of the control parameter.

New insight into the structure of systems at the phase transition point has been obtained by deriving spectral signatures at the critical point of a quantum phase transition (called critical symmetry or critical matter). Experimental examples of critical matter ^{152}Sm and ^{150}Nd have been found. The excitation spectrum of critical matter is at the present time of great interest in condensed matter physics, where opposing views have been presented.[15,16] The behavior discussed in Sect. 3, when extrapolated to infinite systems, is in accord to that of 16 rather than to that of 15.

Acknowledgments

This work was performed in part under DOE Contract No. DE-FG-02-91ER40608. I wish to thank Mark Caprio for help in preparing the manuscript.

References

1. E. Ising, Z. Phys. **31**, 253 (1925).
2. For a review, see F. Iachello and A. Arima, *The Interacting Boson Model*, Cambridge University Press, Cambridge, 1987.
3. D.D. Warner and R.F. Casten, Phys. Rev. C **28**, 1798 (1983).
4. J. Ginocchio and M.W. Kirson, Phys. Rev. Lett. **44**, 1744 (1980).
5. A.E.L. Dieperink, O. Scholten and F. Iachello, Phys. Rev. Lett. **44**, 1747 (1980).
6. R. Gilmore, J. Math. Phys. **20**, 891 (1979).
7. D.H. Feng, R. Gilmore and S.R. Deans, Phys. Rev. C **23**, 1254 (1981).
8. F. Iachello and N.V. Zamfir, Phys. Rev. Lett. **92**, 212501 (2004).
9. A. Leviatan and J.N. Ginocchio, Phys. Rev. Lett. **90**, 212501 (2003).
10. F. Iachello, Phys. Rev. Lett. **85**, 3580 (2000).
11. F. Iachello, Phys. Rev. Lett. **87**, 052502 (2001).
12. A. Bohr, Mat. Fys. Medd. Dan. Vid. Selsk. **26**, No. 14 (1952).
13. R.F. Casten and N.V. Zamfir, Phys. Rev. Lett. **85**, 3584 (2000).
14. R.F. Casten and N.V. Zamfir, Phys. Rev. Lett. **87**, 052503 (2001).
15. B.A. Bernevig, D. Giuliano, and R.B. Laughlin, Ann. Phys. **311**, 182 (2004).
16. T. Senthil *et al.*, Proc. Int. Conf. on *Statistical Physics of Quantum Systems*, in press (2004).

NUCLEAR-STRUCTURE THEORY IN THE SEARCH FOR NEW FUNDAMENTAL PHYSICS

JONATHAN ENGEL

Department of Physics and Astronomy, University of North Carolina, CB3255
Chapel Hill, North Carolina, 27599-3255, United States of America
engelj@physics.unc.edu

Received 1 December 2004

Nuclear-structure theory is an important ingredient in the interpretation of many experiments that look for physics beyond the Standard Model. I review the role of nuclear structure in attempts to to learn more about neutrinos through double beta-decay and to discover new sources of CP violation through atomic electric-dipole moments.

Keywords: Nuclear physics; many-body theory; standard model.

1. Introduction

There is a lot we don't know about the universe. First, there are observed phenomena for which we have no explanation, or at least no coherent and accepted theory. We know that most of the matter in the universe is not made of baryons, but don't know what it is made of. We know that the expansion of the universe is accelerating but have not the slightest clue about the cause. And, as I discuss below, we have recently discovered that neutrinos have mass but remain ignorant of the values and causes of their masses.

There are also things we haven't observed but strongly suspect are present. An example is the existence of new particles at or just above a few hundred GeV (the weak scale). Though we have no observational evidence for these particles, it is hard to understand why the Higgs boson has the mass it does without new physics at that scale.[1] Another example, which I discuss here, is a source of CP violation other than the phase in the CKM matrix that is apparently responsible for observed phenomena in the K and B systems. A new source is required for a natural explanation of the large excess of matter (over antimatter) in the universe.[2] Many theorists suspect, in fact, that the new particles just above the weak scale are responsible for the CP violation. Supersymmetry is the most popular explanation of both phenomena.

Many experiments have been set up to elucidate the new physics. In what follows I discuss two types: searches for neutrinoless double-beta decay, which will provide information about neutrinos, and measurements of atomic-electric dipole moments, which tell us about CP violation. Interpreting the results of both sets of experiments will require a precise understanding of the many-body physics inside atomic nuclei. Here I discuss that physics, assess our current understanding, and outline improvements, in both the short and long terms, that will allow us to better extract information from experimental results.

2. Neutrinos

2.1. *What we know and don't know*

In the last few years, measurements of flavor oscillations in neutrinos produced by the sun and in the atmosphere have led to the conclusion that the three known species of neutrinos mix and have mass. The measured quantities are two differences in the squares of the masses — $\Delta m^2_{\text{sol}} \approx 7 \times 10^{-5}$ eV^2 from solar experiments and $\Delta m^2_{\text{atm}} \approx 2 \times 10^{-3}$ eV^2 from atmospheric experiments — and two corresponding mixing angles: $\theta_{\text{sol}} \approx 33°$ and $\theta_{\text{atm}} \approx 45°$. The atmospheric experiments and experiments with nuclear reactors imply that the third mixing angle θ_{e3}, which measures the electron-neutrino component in the mass eigenstate that participates strongly in atmospheric oscillations but not solar oscillations, is small.

There are several very important things that we don't know: the overall mass scale, i.e. the mass of the lightest neutrino (we know only the differences between the squares of the masses), the pattern in which they are arranged (normal vs. inverted hierarchy — see Fig. 1), and whether neutrinos are their own antiparticles, i.e. whether they are "Majorana particles". A good experiment on neutrinoless double-beta decay will help us address all these questions.[3]

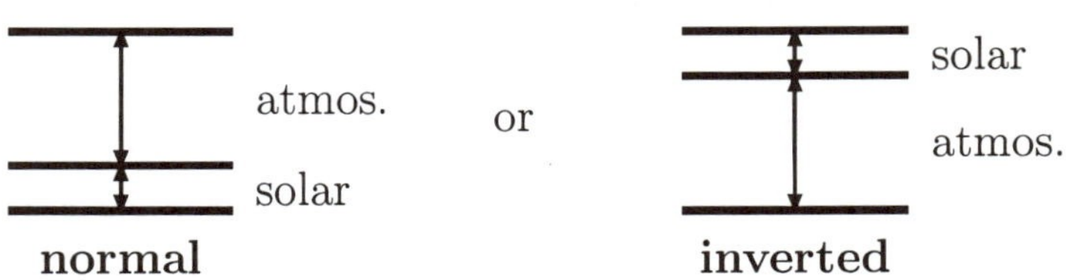

Fig. 1. Two possible arrangements of neutrino masses. In the "normal hierarchy", the two lightest neutrinos are separated by the very small solar-oscillation mass, and the third is significantly heavier. In the "inverted hierarchy" the two lightest are separated by the larger atmospheric mass and the third is very slightly heavier than the second. The drawings are not to scale.

2.2. *Neutrinoless double-beta decay and neutrino properties*

Neutrinoless double-beta decay is a process in which two neutrons decay into two protons, emitting two electrons and a virtual neutrino that is reabsorbed (see Fig. 2). A right-handed antineutrino is emitted from the first vertex and a left-handed neutrino absorbed at the second; for this to happen neutrinos must be Majorana particles. In addition, provided no mechanism other than virtual-neutrino exchange induces the decay, the rate is proportional to the square of an average of neutrino

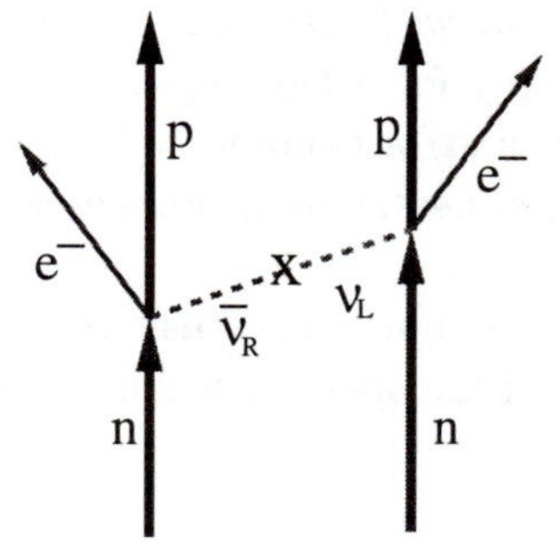

Fig. 2. Feynman diagram of neutrinoless double-beta decay

masses, where each mass in the average is weighted by the degree of mixing of the corresponding eigenstate with the electron neutrino. It is this average mass that is responsible for flipping handedness. If the lightest neutrino is very light than the average is quite different for the two hierarchies discussed above:

$$\langle m \rangle \approx \sqrt{\Delta m^2_{\rm sol}} \sin^2 \theta_{\rm sol} \ \ (\text{normal}) \ , \qquad \langle m \rangle \approx \sqrt{\Delta m^2_{\rm atm}} \cos 2\theta_{\rm sol} \ \ (\text{inverted}) \ . \tag{1}$$

The reason for these different values is that in the first case the heaviest eigenstate contains very little electron neutrino, leading to the appearance only of $\sqrt{\Delta m^2_{\rm sol}}$ in $\langle m \rangle$, while in the second the heaviest eigenstate contains a significant amount of electron neutrino, so that $\sqrt{\Delta m^2_{\rm atm}}$ dominates the expression. The large value for $\langle m \rangle$ means that the next generation of double-beta decay experiments will be sensitive enough in principle to confirm or rule out the inverted hierarchy if neutrinos are Majorana particles. I've inserted the caveat "in principle" because the decay rate depends on a nuclear matrix element as well as on the average neutrino mass. If the matrix element is very different from what we think it is, an unobservably slow decay may not mean that the hierarchy is normal. Accurate knowledge of the matrix element is crucial.

2.3. *Nuclear-structure calculations*

The double-beta matrix element has the approximate form[3]

$$M_{fi} \approx \langle f | \sum_{j,k} \left[\frac{g_V^2 - g_A^2 \vec{\sigma}_j \cdot \vec{\sigma}_k}{|\vec{r}_j - \vec{r}_k|} \right] \tau_j^+ \tau_k^+ | i \rangle \ , \tag{2}$$

where $|i\rangle$ and $|f\rangle$ are the initial and final ground states, g_V and g_A are the weak vector and axial-vector decay constants for the nucleon, the operator τ^+ turns a neutron into a proton, and the sums are over all the neutrons in the initial nucleus. The matrix element, unfortunately, is hard to calculate with an accuracy better than a factor of three or four (which is not good enough). There are several reasons:

- Double-beta decay is allowed energetically (at the same time as single-beta decay is forbidden) only for a few nuclei, most of which are heavy (as in, e.g., ^{136}Xe $\longrightarrow^{136}$Te) and/or complicated (e.g. in ^{76}Ge $\longrightarrow^{76}$Se).
- There are no measured neutrinoless double-beta decays with which to calibrate theory.
- M_{fi} involves a two-body operator, sensitive to delicate spin correlations.
- Most of the strength of the operator is much higher in energy than the final ground state $|f\rangle$.

$\vdots$

Nonetheless, a huge number of calculations have been published, mostly within the "neutron-proton QRPA". This method treats states in the intermediate nucleus as neutron-proton phonons on initial and final ground states. It tends to be applied with quite phenomenological interactions, the parameters of which are adjusted to reproduce measured single-beta strength distributions and/or rates of *two*-neutrino double-beta decay, a related process, in which neutrinos are emitted instead of exchanged, that has been observed in several nuclei. Another more rigorous approach, the nuclear shell model, has also been applied, though much less often. In the shell model, because fewer single-particle levels are involved (though the correlations within them are treated more completely), the effective interaction is obtained more carefully, usually from perturbation theory followed by some tweaking to reproduce spectra.

The results of all these calculations for the decay of ^{76}Ge are summarized in Fig. 3, which plots the number of calculations that would yield a particular neutrino mass $\langle m\rangle$ if the decay lifetime were 4×10^{27} years. There are so many results that a statistical analysis is almost possible. All of the results but one were obtained in the QRPA. The histogram indicates a "statistical" uncertainty of about a factor of three in the value of the neutrino mass that will be extracted from any positive experimental result. Of course the figure doesn't include "systematic" uncertainty, but if we assume that is small we get an idea of where structure theory stands.

Two special results are indicated by arrows. The first is the lone shell-model number (near the heavier-mass side) and the second the conclusion of an analysis[4] that attempted to consolidate the QRPA calculations by making sure they also reproduce the measured two-neutrino decay rate. My unjustifiable suspicion is that the true number lies somewhere between these two arrows.

In order to test this hypothesis, and to really give the experimenters full bang for their buck, we have to improve the calculations. In the short term, the best hope for an accurate matrix element is an improved shell-model calculation. There are two obvious ways to make improvements: larger configuration spaces and renormalization of the double-beta-decay operator in Eq. 2. The first goal can be reached, e.g., through the "factorization" method of Ref. 5. The basic idea, which is similar to that underlying the density-matrix-renormalization group, is to use the singular-value decomposition theorem to write the ground states of the initial and final nuclei

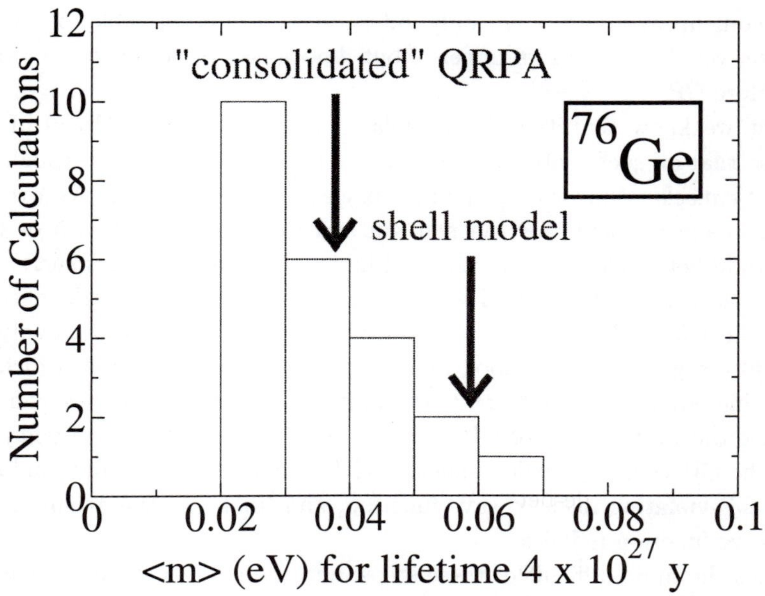

Fig. 3. Collected calculations of the matrix element governing the double beta decay of ^{76}Ge, presented in the form of the average neutrino mass $\langle m \rangle$ that one would deduce from a measured lifetime of 4×10^{27} years.

in terms of proton (p) and neutron (n) states as

$$|0\rangle \equiv \sum_{i,j} \Psi_{i,j} |p_i\rangle |n_j\rangle \longrightarrow \sum_i s_i |\tilde{p}_i\rangle |\tilde{n}_j\rangle \tag{3}$$

and then exploit the rapid falloff of the s_i's to truncate the basis. Constructing an accurate effective decay operator for this space will require more ingenuity, but should also be possible through perturbation theory. My hope is that these steps will allow us to reduce the uncertainty in the matrix element from three or four to one-and-a-half or two.

In the longer term, we hope to exploit *ab initio* approaches that use the bare nucleon-nucleon interaction and are now restricted to very light nuclei. Coupled-clusters techniques, nonperturbative operator renormalization, and perhaps even variational Monte Carlo may some day be applied to double-beta decay.

3. CP Violation

3.1. *Atomic electric dipole moments*

If the CPT theorem is obeyed by nature (so that CP violation implies time-reversal (T) violation) then states whose only degeneracy is a $2J + 1$ spin multiplicity can have electric dipole moments (EDMs) only if CP is violated. The reason, roughly speaking, is that if T is conserved such a moment $\vec{d}$ must be proportional to the angular momentum vector $\vec{J}$, the only independent vector available to a nondegenerate state, yet $\vec{J}$ changes sign under T but $\vec{d}$ does not. So a nonzero $\vec{d}$ violates T, and therefore CP.

In fact we know that the CP is violated in the Standard Model of particle physics, so that all such states do have EDMs, but the violation is through a single phase that cancels in flavor-diagonal matrix elements of operators. As a result, the predicted dipole moments for the neutron and for atoms are tiny, 5 or 6 orders of magnitude below current experimental limits. However, theories such as supersymmetry that try to keep the Higgs boson light, account for dark matter, etc., have several independent phases, and can violate CP much more strongly. EDMs are therefore a great place to look for new physics. Atomic EDMs, which can be measured incredibly precisely[a] are good things to look for provided we can theoretically trace the propagation of CP violation from, e.g. a phase in supersymmetry through the CP-violating nucleon-nucleon (NN) interaction the phase induces, the resulting CP-violating nuclear wave function, and finally to the atomic EDM the nuclear wave function induces.

Perhaps the most difficult of these steps, and the one I focus on, is calculating the effect of a CP-violating (NN) interaction on the nucleus. The nuclear EDM turns out to be irrelevant as far as the atom goes because, as demonstrated by Schiff,[7] the electrons rearrange themselves to cancel it. Instead, the atomic dipole moment is proportional to a residual nuclear dipole-like quantity, commonly referred to as the "Schiff moment", given by the expectation value of the operator

$$\vec{S} = \sum_p (r_p^2 - \frac{5}{3}\langle R^2 \rangle)\vec{r}_p \ , \tag{4}$$

where the sum is over protons and $\langle R^2 \rangle$ is the mean square charge radius.

The CP-violating interaction H_{CP} responsible for inducing the Schiff moment can be written approximately as

$$\hat{H}_{CP} = -\frac{g}{4\pi m_N} \left\{ [(\bar{g}_0 - \bar{g}_2)\, \boldsymbol{\tau}_1 \cdot \boldsymbol{\tau}_2 + 3\,\bar{g}_2\, \tau_{1z}\tau_{2z}]\, (\vec{\sigma}_1 - \vec{\sigma}_2) + \bar{g}_1\, (\vec{\sigma}_1\tau_{1z} - \vec{\sigma}_2\tau_{2z}) \right\}$$
$$\cdot (\vec{\nabla}_1 - \vec{\nabla}_2) \left(\frac{\exp(-m_\pi |\vec{r}_1 - \vec{r}_2|)}{|\vec{r}_1 - \vec{r}_2|} \right), \tag{5}$$

where τ_z is +1 for protons and -1 for neutrons, m_N is the nucleon mass, and g is the strong pion-nucleon coupling constant. The interaction also involves three unknown dimensionless parameters $\bar{g}_i$, the values of which depend on the underlying source of CP violation. The task of nuclear structure is to evaluate the dependence of the Schiff moment on these parameters.

---

[a]Though a nonzero EDM has been seen yet, the upper limit on that of ^{199}Hg is about 2×10^{-28} e cm.,[6]

Because the CP-violating interaction is so weak it can be treated in perturbation theory, so that the Schiff moment is given by

$$\langle \vec{S} \rangle = \sum_m \frac{\langle 0|\vec{S}|m\rangle\langle m|H_{CP}|0\rangle}{E_0 - E_m} + c.c. \tag{6}$$

where the $|m\rangle$ are excited states. A good calculation therefore must produce the strength distribution to excited states of the Schiff operator. Fortunately, although this distribution has not been measured, that of the isoscalar version of the Schiff operator has been, at least in ^{208}Pb, through scattering of alpha particles. We can therefore test a calculation to by looking at its predictions for the isoscalar distribution in lead.

Recently, it was realized that in pair-shaped nuclei (of which there are not many) Schiff moments are enhanced by factors of 100 or more. The reason is that the ground state has a very low-lying partner with the same angular momentum and opposite parity that is excited strongly by the Schiff operator. In such nuclei — ^{225}Ra is a prominent example — we need only calculate the Schiff strength to that one state, not the entire strength distribution. Since the excited state has essentially the same intrinsic structure as the ground state, we can approximately factor out the parity and orientation dependence and calculate intrinsic-ground-state expectation values of $\vec{S}$ and H_{CP}.

3.2. *Calculating Schiff moments*

Schiff moments are proportional to the nuclear charge and so are largest in very heavy nuclei. For such systems the state of the theoretical art is self-consistent "Skyrme" mean-field theory and RPA with a completely phenomenological zero-range force fit to ground-state properties and spin excitations. Figure 4 shows the predictions of one of these forces for the ground-state intrinsic densities of the even Radium isotopes. A transition from spherical shapes through pear shapes to symmetrically deformed shapes is apparent; ^{225}Ra, which is particularly useful for an experiment, has 137 neutrons and is quite "pear-shaped".

Table 1 shows the predictions of several Skyrme interactions and the simple particle-rotor model for quantities that are related to the Schiff moment. The second column contains the intrinsic Schiff moment associated with the deformed shape, and the third and fourth the intrinsic-state expectation values of operators that are contained in H_{CP} in the limit of very heavy pions (a simplifying limit). These matrix elements are considerably smaller in the self-consistent calculations than in the particle-rotor model, where Schiff moments of pear-shaped nuclei were first examined.[8]

3.3. *Recent interpretation of Skyrme-mean-field theory*

The parameters in Skyrme interactions are fit to reproduce ground-state nuclear properties. This fact and the ability to represent Skyrme interactions as an energy-

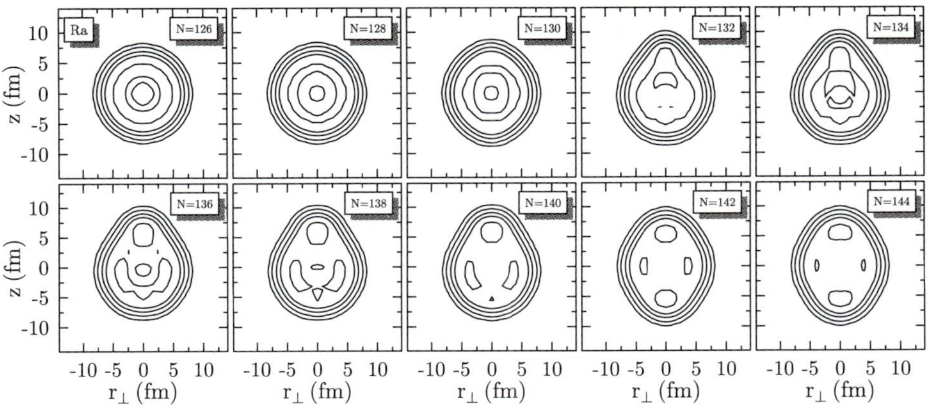

Fig. 4. Ground-state intrinsic densities in the Radium isotopes predicted by a Skyrme interaction. Here $r_\perp = \sqrt{x^2 + y^2}$.

Table 1. Intrinsic-state Schiff moment in ^{225}Ra, and expectation values of important matrix elements. The first column contains a set of commonly used Skyrme interactions, the second the intrinsic-state Schiff moment, and the third and fourth the intrinsic-state integrals of the neutron spin density against the gradients of the overall neutron and proton densities.

	$\langle S_z^{\rm int}\rangle$	$\langle \vec{\sigma}_n \cdot \vec{\nabla}\rho_n\rangle$	$\langle \vec{\sigma}_n \cdot \vec{\nabla}\rho_p\rangle$
SIII	35	−0.58	−0.49
SkM*	47	−0.62	−0.12
SLy4	43	−0.63	−0.05
SkO'	42	−0.32	−0.11
part./rot.	24	−1.20	−0.80

density functional has led recently to the statement that calculations like that just described include correlations beyond mean-field theory, that what we call Hartree-Fock equations are really Kohn-Sham equations with an approximate density functional. How justified is this statement?

The answer is not obvious. Nuclei are self-bound systems without external potentials, and no rigorous Kohn-Sham theory exists for such systems. In addition, the densities we care about are intrinsic densities, measured with respect to the nuclear center of mass, not lab-frame densities, which are spread out uniformly over all space. The intrinsic density is represented by an N-body operator, in sharp contrast with the normal one-body density operator. These problems can be overcome and an intrinsic-density functional theory formulated; perhaps the functional can even be derived, though progress toward that goal is slow[b] Skyrme functionals, however, are also applied to deformed nuclei, in which the intrinsic density is not spherically symmetric, as in Fig. 4. Unfortunately, unlike center-of-mass motion, rotational motion does not decouple from intrinsic-frame motion; indeed the oriented intrinsic density is not a real observable because of an ambiguity in defining principal axes. The question of whether we can include arbitrary correlations in a Kohn-Sham-like calculation with deformed orbitals is completely open. Though such calculations are becoming more accurate through more systematic fitting to data, dramatic improvements will require serious attempts to understand the limits of nuclear density-functional theory and calculate the functionals from first principles.

4. Other New Physics

There are may other problems in particle physics and cosmology where nuclear-structure theory plays a role in the interpretation of experiments. Just a few examples: The scattering cross sections of dark-matter particles from detectors on earth, besides depending on the nature of the particles, depend on the structure of the nuclei the particles hit in the detector. Corrections to the standard V-A weak interactions are searched for in single-beta decay, where nuclear structure plays an obvious role. Nucleosynthesis in supernovae (and the explosions themselves) can tell us about neutrino properties but only if we understand nuclear physics at unusual densities. As I hope I've conveyed, nuclear-structure theory in the heavy nuclei in which many of these processes take place is still in large part an art. Our job is to make it more precise.

Acknowledgments

This work was supported in part by the U.S. Department of Energy under grant DE-FG02-97ER41019.

References

1. L. Susskind, *Phys. Rev.* **D20**, 2619 (1979).
2. See, e.g., W. Bernreuther, *Lect.Notes Phys.* **591**, 237 (2002).
3. For a recent review, see S.R. Elliott and J. Engel, *Journal of Physics G: Nucl. Part. Phys.* **30**, R183 (2004).
4. V.A. Rodin, A. Faessler, F. Simkovic, and P. Vogel *Phys. Rev.* **C68**, 034016 (2003).
5. T. Papenbrock, A. Juodagalvis, and D.J. Dean, *Phys. Rev.* **C69**, 024312 (2003).
6. M.V. Romalis, W.C. Griffith, J.P. Jacobs, and E.N. Fortson, *Phys. Rev. Lett.* **86**, 2505 (2001).
7. L.I. Schiff,*Phys. Rev.* **132**, 2194 (1963).
8. V. Spevak, N. Auerbach, and V. V. Flambaum, *Phys. Rev.* **C 56**, 1357 (1997).
9. K.T.R. Davies, R. J. McCarthy, J.W. Negele, and P.U. Sauer, *Phys. Rev.* **C10**, 2607 (1974).

---

[b]The only convincingly derived nuclear density functional includes a limited set of correlations.[9]

MATTER AT EXTREME DENSITY AND ITS ROLE IN NEUTRON STARS AND SUPERNOVA

SANJAY REDDY

Theoretical Division, Los Alamos National Laboratory, New Mexico 87545, USA
reddy@lanl.gov

Received 1 December 2004

We discuss observable aspects of neutron stars and supernova that are influenced by the properties of matter at extreme density. In particular, we explore the possible role phase transitions to quark matter phases at supra nuclear density. The competition between the strange quark mass and the pairing energy in quark matter, and show that it leads to a rich phase structure at densities of relevance to neutron stars. The equation of state and transport properties of quark matter is shown to be strongly influenced by pairing correlations at the Fermi surface.

Keywords: Neutron stars; supernova; dense matter; superconductivity.

1. Introduction

Since their discovery, neutron stars have fascinated the many body theorist. These compact objects provide a unique laboratory to probe matter at the most extreme density. The central density of a neutron stars is expected to be $5 - 10\ \rho_0$, where $\rho_0 = 0.16$ baryons/fm^3 is the nuclear saturation density. During their birth in the aftermath of a core-collapse supernova explosion, the protoneutron star is also characterized by extreme temperatures. Simulations indicate $T \sim 50$ MeV.[1] In this talk we discuss how observable aspects of neutron stars are affected by physics of the inner core. In particular, we explore the role of phase transitions to deconfined quark phase where pairing correlation at the Fermi surface are likely to be important. We highlight recent work on the phase structure, the equation of state, and transport properties in superconducting quark matter.

We begin by discussing observable aspects of neutron star structure in Sec. 2. This is followed by a brief review of recent progress in our understanding of the phase structure of quark matter in Sec. 3. The quark matter equation of state (EoS) and its relation to the structure of compact stars is discussed in Sec. 4. Transport properties, in particular the neutrino mean free path in superconducting phases of quark matter and its relevance in the early thermal evolution is addressed in Sec. 5. A detailed discussion of the role of novel high density phases on the structure and evolution of neutron stars can be found in Ref. 2.

2. Observable Aspects: Neutron Star Structure and Thermal Evolution

The masses of neutron stars in the binary system 1913+16 is perhaps one of the most accurately determined quantities in astronomy.[3] Table 1, lists the masses of several neutron stars. This list, which is a subset of the 25 measured masses, consists of neutron stars in three distinct astrophysical systems: 1) neutron star binaries, 2) white dwarf-neutron star systems, and 3) x-ray binaries. Masses in the double neutron star system are the most accurately known and appear to cluster in a narrow interval centered at 1.35 $M_\odot$. In the other systems, the distribution is broader (albeit substantially larger errors) and seem to indicate the existence of heavier neutron stars. The Vela pulsar and 4U1700-37 have masses similar to 2 $M_\odot$. However, the inherent uncertainty in these systems is large. For example the object 4U1700-37 with $M \simeq 2.4 \pm 0.3\ M_\odot$ is yet to be uniquely identified as a neutron star (it could be a low mass black-hole) b.[4] Further, there are several potential sources of unknown systematic error in these systems.[5] Notwithstanding these caveats, the conservative lower bound on the mass of Vela-X1 of $M_{\rm min} \simeq 1.74 M_\odot$[5] suggests that the clustering at low mass (1.35 $M_\odot$) seen in double neutron stars might be specific to those systems and may not be related to the intrinsic limiting mass. In addition, recent analysis of neutron star masses in neutron star-white dwarf binaries also strengthens the case for neutron stars heavier than the canonical value of 1.4 $M_\odot$.[6]

Star	Mass ($M_\odot$)	type	Star	Mass ($M_\odot$)	type
1913+16	1.4411(4)	ns-ns	1913+16 (c)	1.3874(4)	ns-ns
B1534+12	1.339(3)	ns-ns	B1534+12 (c)	1.339(3)	ns-ns
B2127+11C	1.349(40)	ns-ns	B2127+11C (c)	1.363(40)	ns-ns
J1713+0747	1.45(31)	wd-ns	J0751+1807	2.20(40)	wd-ns
J0621+1002	1.70(30)	wd-ns	J0437-4715	1.60(20)	wd-ns
Vela X-1	1.88(13) - 2.27(17)	xrb	4U1700-37	2.4(3)	xrb

Unlike masses, radii of neutron stars are poorly known. The recent discovery of a nearby isolated neutron star RXJ185635 is perhaps the most promising candidate for a radius measurement.[7] X-ray emission from the neutron star has been observed by the Chandra and ROSAT x-ray satellites and the optical counter part has been measured by the Hubble space telescope. The x-ray observations are well described by a black body with a temperature of 63 eV. For a black-body emitter, the total flux at a large distance $d \gg R_S$ is $F = \sigma_{\rm sb} T_\infty^4\ R_\infty^2/d^2$, where $T_\infty = T\sqrt{1 - R_S/R}$ and $R_\infty = R/\sqrt{1 - R_S/R}$ and $R_S = 2GM_\odot/c^2$ is the Schawrzchild radius . Using the black body fit to the x-ray data from Chandra, one obtains $R_\infty/d \simeq 0.037$ km pc^{-1}. The distance to the RXJ185635 is $d = 117 \pm 12$ pc - implying $R_\infty \simeq 4.2$ km. This is significantly smaller than the canonical radius (10 km) and prompted speculation that RXJ185635 may be a self bound strange quark star.[8] However, the optical data is inconsistent with the black body assumption.[9] A naive extrapolation to optical frequency of the x-ray data significantly under predicts the optical flux - suggesting either a non-uniform surface temperature or the presence of a atmosphere. Theoretical attempts to model both these possibilities are still very rudimentary and do not provide an satisfactory description of the data. Nonetheless, the best fit heavy element atmosphere model indicates that the radius is significantly larger than 4.2 km. It favors $R_\infty \simeq 15 \pm 3$ km and red-shift $z \sim 0.35 \pm 0.15$ which in turn imply that the radius $R \simeq 11 \pm 2$ km and a mass $M \simeq 1.7 \pm 0.4 M_\odot$.[9]

The long term thermal evolution of neutron stars is inferred from observations of x-ray luminosity from neutron stars whose ages can be estimated. The rate of cooling provides an independent means of probing the interior. The surface temperature of aging neutron stars is strongly coupled to the temperature of the core. The high density core is characterized by temperatures which are about 10^{10} K shortly after birth. The cooling the ensues during the first million years, is driven by neutrino emission from the core of the neutron star. The large density and temperature of the core compensates for the weak coupling between neutrino and matter. The dominant neutrino cooling process depends sensitively on the composition and phase structure of the dense core (ror recent reviews, see articles by Yakovlev and Pethick[12] and Page et al.[13])

The early evolution of the proto-neutron star born in the aftermath of the supernova explosion can be inferred from the observation of supernova neutrinos. Neutrinos play an important role in stellar evolution. By virtue of their weak interactions with matter they provide a mechanism for energy loss from the dense stellar interiors. Neutrinos emitted from the proto-neutron star can be detected in terrestrial detectors such as Super Kamiokande and SNO. Current estimates predict $\sim 10,000$ events in Super Kamiokande and ~ 1000 events in SNO from a supernova at the center of our galaxy[11] (distance=8.5 kpc) . These neutrinos are the only direct probes of both the dynamics of gravitational collapse and the properties of dense matter inside the newly born neutron star. The emission time scale is set by neutrino diffusion. Hence the duration of the neutrino signal in terrestrial detectors is intimately connected with the neutrino opacity of matter deep inside the neutron star.

3. Phase Structure of Dense Quark Matter

The ground state of QCD at large densities is likely to be a color superconductor.[14,15,16,17] When the quark chemical potential μ is very large compared to the strange quark mass m_s, three flavor quark matter is expected to be in a particularly symmetric phase called the Color-Flavor Locked (CFL) phase, with BCS pairing involving all nine quarks.[19] In this phase the quark excitation spectrum has a gap Δ and model calculations indicate that $\Delta \sim 50 - 100$ MeV for $\mu \sim 400$ MeV. This phase breaks the $SU(3)_{\rm color} \times SU(3)_L \times SU(3)_R \times U(1)_B$

symmetry of QCD down to the global diagonal $SU(3)$ symmetry. The lightest excitations are an octet of pseudo-Goldstone bosons and a true Goldstone boson associated with the breaking of the global $U(1)_B$ symmetry. For μ not much larger than m_s, however, the stress induced by the strange quark mass on the symmetric CFL state might not be negligible: less symmetric phases are possible. One possibility is the CFLK^0 phase, wherein a condensate of K^0 mesons exists in addition to the diquark condensate of the CFL phase–this phase breaks hypercharge and isospin symmetries.[21,22] Another possibility is the LOFF phase, which exhibits crystalline color superconductivity–the diquark condensate varies periodically in space, breaking translation and rotation symmetries.[20] Most recently, a superconducting phase of two [23] and three [24] flavor quark matter with non-trivial gapless fermionic excitations has been suggested . However, the stability of these gapless phase is currently not well understood. In particular, it has been shown that the Meissner masses of some of gluons in these gapless phases can become negative.[25] In the two flavor case, this can be shown to imply that the homogeneous gapless phase is unstable to the formation of a mixed heterogeneous phase where both normal and superconducting phases coexist.[26]

In compact objects, where $\mu \sim 400$ MeV is not much large than m_S the phase structure is less clear. Were quark matter occurs at these densities, the stresses induced by m_s could play a dominant role in determining the pairing structure (since $m_s^2/\mu\Delta \sim 1$). In normal quark matter, a large strange quark mass results in the depletion of strange quarks. This in turn requires a splitting between the up and down quark Fermi levels so as to ensure electric charge neutrality.[27] Recent analysis of the role of charge neutrality indicate that two flavor u-d superconductor (2SC) is disfavored relative to the CFL phase.[27,28] However, at the present time since we cannot predict either Δ or m_s with confidence, we must entertain all these aforementioned possibilities. The role of the strange quark mass in determining the phase structure for densities of relevance to neutron stars continues to be an important unanswered question.

4. Mass/Radius and the Equation of State

Fig. 1 shows the mass-radius curve of neutron stars constructed using different equation of states. In the left panel, the curve labeled "Mean-Field" corresponds to the model and the curve labeled APR is a microscopic many-body calculation, by Akmal, Pandharipande and Ravenhall, of the EoS using realistic nucleon-nucleon potentials.[29] The dashed lines labeled $R = R_s$ and $R = 1.52R_s$ (where $R_s = 2GM/c^2$ is Schwarzchild radius) are model-independent constraints that require all stable stars to be to their right. To the left of the line defined by $R = R_s$ the objects must be black holes since the event horizon is outside the object, to the left of the line defined by $R = 1.52R_s$ the star requires a very stiff EoS for which the speed of sound exceeds the speed of light. The last stable orbit around a neutron star is located at a radius $R = 3R_s$. In accreting neutron star systems called QPO's

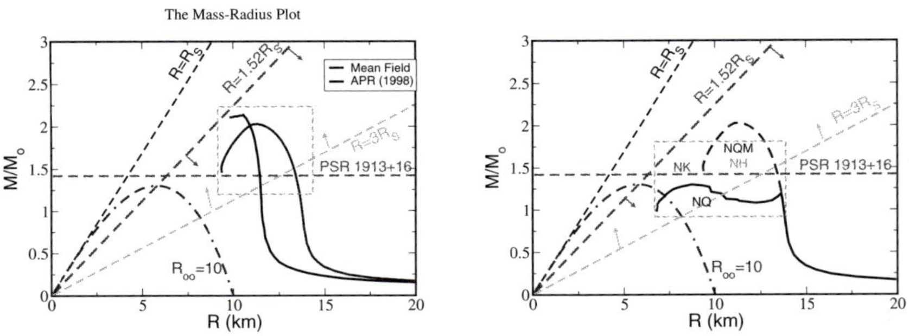

Fig. 1. Left Panel: Mass-radius relation for nuclear stars, where $M_\odot$ is the mass of our sun. Right Panel: Mass-radius relation for several exotic stars. The labels NQ, NQM, NH, and NK correspond to hybrid stars containing normal quark matter, nuclear-quark mixed phases, hyperonic matter and kaon condensates, respectively.

(quasi-periodic oscillators) there is observational evidence of the existence of such orbits. This constrains stars to lie to the left of the line defined by $R = 3R_s$. Accurate measurement of the neutron star mass in the PSR1913+16 introduces a further constraint which requires the maximum mass of the stars constructed using any model EoS be at least as large as 1.44 $M_\odot$ (Table 1 indicates the presence of heavier stars, but the masses in these systems have potential sources of systematics unknown systematic error). Finally, an accurate determination of R_∞ would greatly constrain the allowed region in the mass radius diagram. For example, a precise measurement of an R_∞ of 10 km constrains all stars to lie on the curve labeled $R_\infty = 10$ km in Fig. 1.

Phase transitions at high density generically result in an increase in the energy density of matter at a given pressure. We will loosely refer to this as softening. An EoS whose energy density is on-average large compared to the nuclear EoS can be considered to be soft. A larger energy density increases the gravitation attraction inside the compact object. Consequently, it will require a higher central pressure to achieve hydrostatic equilibrium. However, with increasing pressure, the energy density increases rapidly in a soft EoS. The competition between matter pressure and gravity is heavily stacked in favor of gravity in a stars with a soft EoS. This results in lower maximum masses and smaller radii. This trend is shown Fig. 1.

The M-R relationship for hybrid stars containing nuclear matter at low density and pure quark matter at high density is shown (black curve labeled NQ). Other exotic possibilities like hyperons (NH), kaon condensation (NK) and phase transitions in which a mixed nuclear-quark phase is favored (NQM) have a similar effect on the mass-radius curve. Kaon condensation results in stars that are relatively compact while the hyperonic and the nuclear-quark mixed phase stars favor modestly larger maximum masses and radii. For a detailed discussion of how color-superconductivity affects the M-R of neutron stars see Ref. 30. A rough estimate of the region of the

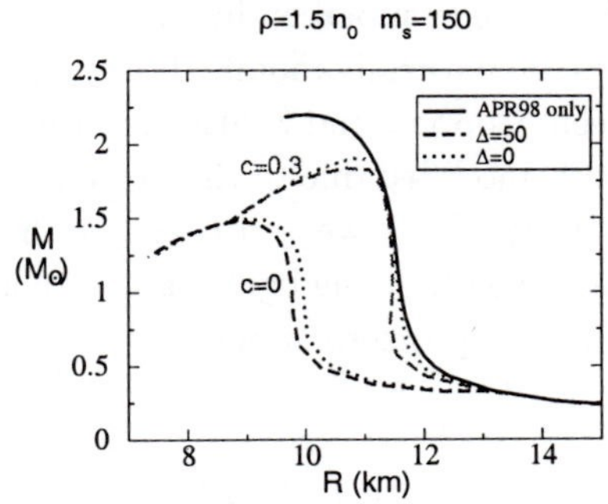

Fig. 2. Mass-Radius of hybrid stars with quark matter cores. The transition density was fixed at $1.5\rho_0$.

M-R diagram which can be populated by exotic stars given the uncertainties in these models is indicated by the dashed box. While phase transitions typically result in a softer EoS, there are exceptions. For example, consider a Bag - model inspired quark matter EoS of the form

$$P(\mu) = (1-c)\,\frac{3}{4\pi^2}\mu^4 - \frac{3}{4\pi^2}m_s^2\mu^2 + \frac{3}{\pi^2}\Delta^2\mu^2 - B\,, \tag{1}$$

where m_s, Δ and B are the strange quark mass, the pairing energy and the effective bag constant, respectively and c is a parameter the mocks up corrections to the kinetic energy term arising due to strong interactions. For a chosen transition density, the mass-radius is found to be sensitive to the value c. This is shown in Fig.2. For $c = 0.3$ and a transition density in the range $\rho_0 - 2\rho_0$ the radius of neutron stars in the mass range $1 - 1.5M_\odot$ closely mimics the behavior seen in nuclear stars.[31]

Mass and radius are integral quantities and are sensitive to the EoS over a wide range of densities. Therefore, in principle one requires a large number of different and simultaneous mass-radius measurements to directly infer the EoS. However, our knowledge of the nuclear properties at nuclear saturation density and nuclei provide valuable additional information. This is encoded in most realistic equations of state and lead to the generic trends described above. The accurate measurement of both the mass and radius of star will provide a stringent constraint on the high density EoS. For example, the discovery of a star with a $R \lesssim 8$ km would provide strong support for the existence of novel phases inside neutron stars. On the other hand, the accurate measurement of neutron star mass $M \gtrsim 1.8\ M_\odot$ would support the absence of phase transitions. While neither of these hypothetical measurements will provide conclusive evidence, they will provide valuable guidance to the theoretical models of high-density matter.

5. Transport Properties

As discussed earlier, the rate of production and propagation of neutrinos in the dense interior of the neutron star affects its thermal evolution. In this section, we address how neutrinos propagate in superconducting and superfluid quark matter.[32] The free quark propagators are naturally modified in a superconducting medium. As first pointed out by Bardeen, Cooper, and Schrieffer several decades ago, the quasi-particle dispersion relation is modified due to the presence of a gap in the excitation spectrum. In calculating these effects we will consider the simplified case of QCD with two light quark flavors given that the light u and d quarks dominate low-energy phenomena. In this two flavor, spin zero superconductor (2SC) the anomalous (or Gorkov) propagator is

$$F(p)_{abfg} = -i\epsilon_{ab3}\epsilon_{fg}\Delta \left(\frac{\Lambda^+(p)}{p_o^2 - \xi_p^2} + \frac{\Lambda^-(p)}{p_o^2 - \bar{\xi}_p^2} \right) \gamma_5\, C\,. \tag{2}$$

Here, a, b are color indices, f, g are flavor indices, ϵ_{abc} is the usual anti-symmetric tensor and we have conventionally chosen 3 to be the condensate color. This propagator is antisymmetric in flavor and spin, with $C = -i\gamma_0\gamma_2$. The normal quasi-particle propagators are given as

$$S(p)_{af}^{bg} = i\delta_a^b\delta_f^g \left(\frac{\Lambda^+(p)}{p_o^2 - \xi_p^2} + \frac{\Lambda^-(p)}{p_o^2 - \bar{\xi}_p^2} \right) (p_\mu\gamma^\mu - \mu\gamma_0)\,. \tag{3}$$

This is written in terms of the particle and anti-particle projection operators $\Lambda^+(p)$ and $\Lambda^-(p)$ respectively. The quasi-particle energy is $\xi_p = \sqrt{(|\vec{p}| - \mu)^2 + \Delta^2}$, and for the anti-particle $\bar{\xi}_p = \sqrt{(|\vec{p}| + \mu)^2 + \Delta^2}$.

The appearance of an anomalous propagator in the superconducting phase indicates that the polarization tensor gets contributions from both the normal quasi-particle propagators Eq.3 and the anomalous propagator Eq.2 and is given by

$$\Pi_{\alpha\beta}(q) = -2i\int \frac{d^4p}{(2\pi)^4} \mathrm{Tr}\, \left[S(p)\Gamma_\alpha S(p+q)\Gamma_\beta + F(p)\Gamma_\alpha \bar{F}(p+q)\Gamma_\beta \right]. \tag{4}$$

We can compute the neutrino mean free paths within the BCS approximation since the scattering cross section $d\sigma \propto \mathrm{Im}\Pi_{\alpha\beta}(q)$. The results of Ref. 32 indicate that with increasing Δ, the gap in the quasi-particle excitation spectrum results in an exponential attenuation of the scattering response and corresponding exponential enhancement in the mean free path.

The preceding section assumed that there were no low-energy collective excitations. This is true in the two-flavor superconducting phase of quark matter. For three flavors and when the strange quark mass is negligible compared to the chemical potential the ground state is characterized by pairing that involves all nine quarks in a pattern that locks flavor and color.[19] Diquark condensation in the color-flavor locked (CFL) phase breaks both baryon number and chiral symmetries. The Goldstone bosons that arise as consequence introduce a low-lying collective excitations to the otherwise rigid CFL state. Thus, unlike in the normal phase, where quark excitations near the Fermi surface provide the dominant contribution to the weak interaction rates, in the CFL phase, the dynamics of the low-energy collective states— the Goldstone bosons are relevant.[34] The massless Goldstone boson associated with spontaneous breaking of $U(1)_B$ (baryon number) couples to the weak neutral current. We should expect this mode to dominate the response because the psuedo-Goldstone modes arising from chiral symmetry breaking have non-zero masses.

To incorporate the contribution of the Goldstone mode in the polarization tensor discussed in Eqn.4, we need to include vertex corrections. These corrections, which are necessary to satisfy current conservation and the associated Ward identities, account for the difference between the coupling of neutrinos to the quasi-quarks described by the Nambu-Gorkov propagators and the bare quarks in the normal phase. In recent work with Joydip Kundu, we studied these vertex corrections and structure of the response to neutrinos at finite temperature.[33] Our findings indicate that expect for temperatures close to the critical temperature, an low energy effective field theory (EFT) describing the Goldstone mode provides an excellent description of the low energy response. In the low energy EFT description, neutrinos scatter by either emitting or absorbing these massless Goldstone modes.[34]

Neutral-current processes like $\nu + H \to \nu$ and $\nu \to H + \nu$ dominate the opacity. They are not suppressed by gap in the quark excitation spectrum. For example, the neutrino mean free path due to the reaction $\nu \to H\nu$ can be calculated analytically and is given by

$$\frac{1}{\lambda_{\nu\to H\nu}(E_\nu)} = \frac{256}{45\pi}\left[\frac{v(1-v)^2(4+v)}{4\,(1+v)^2}\right] G_F^2\, f_H^2\, E_\nu^3\,, \tag{5}$$

where the decay constant for the $U(1)_B$ Goldstone boson has been computed in earlier work and is given by $f_H^2 = 3\mu^2/(8\pi^2)$. The neutrino mean free path in the CFL phase are similar to those in the normal quark phase both numerically and parametrically.[34] It is interesting to note that the existence of one massless mode compensates for the large gap in the particle-hole excitation spectrum. To get a sense of how these mean free paths compare with mean free paths in other phase we employ earlier estimates of neutrino mean free path in nuclear matter obtained by Reddy, Prakash and Lattimer[35] and in unpaired quark matter obtained by Iwamoto.[36] The quark chemical potential $\mu = 400$ MeV corresponds to a baryon density $n_B \sim 5n_0$ in quark matter, where $n_0 = 0.16$ fm^{-3} is the nuclear saturation density. In the table below, we compare mean free path of thermal neutrinos ($E_\nu = \pi T$) in these different phases at $n_B = 5n_0$.

phase	process	λ(T=5 MeV)	λ(T=30 MeV)
Nuclear Matter	$\nu n \to \nu n$	200 m	1 cm
	$\nu_e n \to e^- p$	2 m	4 cm
Unpaired Quarks	$\nu q \to \nu q$	350 m	1.6 m
	$\nu d \to e^- u$	120 m	4 m
CFL	λ_{3B}	100 m	70 cm
	$\nu\phi \to \nu\phi$	>10 km	4 m

The findings presented in the table above indicate that neutrino mean free path in CFL matter is similar to or shorter than that in unpaired quark matter in the temperature range $T = 1-30$ MeV. This, surprising, result arises solely due to the novel processes involving Cherenkov absorption and radiation of CFL Goldstone bosons. In contrast, the mean free path in quark matter differs significantly from those in nuclear matter under similar conditions and could potentially lead to differences in supernova neutrino emission time scales.

6. Conclusions

Micro-physical properties of matter at supra-nuclear density is shown to play a role in observable aspects of neutron stars. These observations which have improved greatly due to advances in x-ray and neutrino astronomy hold much promise for the future. Although the possibility of a phase transition to quark matter at high density have been discussed extensively in the past, it is only in recent years that we have come to appreciate the role of pairing in quark matter and its influence on the EoS and transport properties. The structure of compact stars which is influenced by the high density EoS could mimic behavior seen in purely nuclear models resulting in a mass-radius relation that resembles typical nuclear predictions. On the other hand, we found the transport properties were very sensitive to the pairing correlations in quark matter. We discussed in detail the importance of Goldstone excitations in determining the neutrino propagation in CFL matter. We hope that the few examples that we discussed here illustrate the importance of many body theory in modeling observable phenomena associated with compact stars. The phase structure of these objects remains a mystery. Observations continue to provide better clues, ultimately it is improved many body theory combined with astrophysical simulations that will help us identify the state of the densest (observable) matter in the universe.

Acknowledgments

It is a pleasure to thank friends and collaborators: M. Alford, G. Carter, D. Kaplan, J. Kundu, J. Lattimer, M. Paris J. Pons, M. Prakash, K. Rajagopal, M. Sadzikowski, A. Steiner, and M. Tachibana. This work is supported in part by funds provided by the U.S. Department of Energy (D.O.E.) D.O.E. contract W-7405-ENG-36.

References

1. J. A. Pons, S. Reddy, M. Prakash, J. M. Lattimer and J. A. Miralles, Astrophys. J. **513**, 780 (1999);W. Keil and H. T. Janka, Astron. Astrophys. **296**, 145 (1995);A. Burrows and J. M. Lattimer, Astrophys. J. **307**, 178 (1986).
2. S. Reddy, Acta Phys. Polon. B **33**, 4101 (2002)
3. R. A. Hulse and J. H. Taylor, ApJ, bf195, L51 (1975)
4. G. E. Brown, J. C. Weingartner, R. A. M. J. Wijers , ApJ, **463**, 297 (1997)
5. H. Quaintrell et al., Astron. Astrophys. **401**, 313 (2003)

6. D. J. Nice, E. M. Splaver and I. H. Stairs, (2004) arXiv:astro-ph/0411207.
7. F. M. Walter, S. J. Wolk, and R. Neuhäuser, Nature, 389, 358 (1996)
8. J. J. Drake *et al.*, Astrophys. J. **572**, 996 (2002) [arXiv:astro-ph/0204159].
9. J. A. Pons, F. M. Walter, J. M. Lattimer, M. Prakash, R. Neuhauser and P. h. An, Astrophys. J. **564**, 981 (2002); F. M. Walter and J. Lattimer, Astrophys. J. **576**, L145 (2002)
10. M. Burgay *et al.*, Nature **426**, 531 (2003)
11. A. Burrows, D. Klein and R. Gandhi, 1992 Nucl. Phys. Proc. Suppl. **31**, 408 (1992)
12. D. G. Yakovlev and C. J. Pethick, Ann. Rev. Astron. Astrophys. **42**,169 (2004)
13. D. Page, J. M. Lattimer, M. Prakash and A. W. Steiner, (2004) arXiv:astro-ph/0403657.
14. B. C. Barrois, Nucl. Phys. B**129**, 390 (1977)
15. D. Bailin and A. Love, Phys. Rept. **107**, 325 (1984)
16. M. G. Alford, K. Rajagopal, and F. Wilczek, Phys. Lett. B**422**, 247 (1998)
17. R. Rapp, et al., Phys. Rev. Lett. **81**, 53 (1998)
18. K. Rajagopal and F. Wilczek, (2001) [arXiv:hep-ph/0011333]; M. Alford, Ann. Rev. Nucl. Part. Sci. 51, 131 (2001); T. Schafer, (2003) arXiv:hep-ph/0304281. D. H. Rischke, 2004 Prog. Part. Nucl. Phys. 52, 197, (2004);
19. M. G. Alford, K. Rajagopal, and F. Wilczek, Nucl. Phys. B**537**, 443 (1999)
20. M. G. Alford, J. A. Bowers, and K. Rajagopal, Phys. Rev. D**63**, 074016 (2001)
21. P. F. Bedaque and T. Schafer, Nucl. Phys. A**697**, 802 (2002)
22. D. B. Kaplan, and S. Reddy, Phys. Rev. D **65**, 054042 (2002)
23. I. Shovkovy, and M. Huang, , Phys. Lett., B**564**, 205 (2003)
24. M. Alford, C. Kouvaris and K. Rajagopal, Phys. Rev. Lett. **92**, 222001 (2004)
25. M. Huang and I. A. Shovkovy, arXiv:hep-ph/0407049.
26. S. Reddy and G. Rupak, arXiv:nucl-th/0405054.
27. M. Alford, and K. Rajagopal, JHEP **0206**, 031, (2002)
28. A. Steiner, S. Reddy, and M. Prakash, Phys. Rev. D 66, 094007 (2002)
29. A. Akmal, V. R. Pandharipande, and D. G. Ravenhall, *Phys. Rev.* C**58**, 1804 (1998)
30. M. Alford, and S. Reddy, 2003, *Phys.Rev.* D**67**, 074024 (2003)
31. M. Alford, M. Braby, M. Paris and S. Reddy, (2004) arXiv:nucl-th/0411016.
32. G. W. Carter, and S. Reddy, *Phys. Rev.* D**62**, 103002 (2000)
33. J. Kundu and S. Reddy, (2004) arXiv:nucl-th/0405055.
34. S. Reddy, M. Sadzikowski and M. Tachibana, *Nucl. Phys.* A**721**, 309 (2002)
35. S. Reddy, M. Prakash and J. M. Lattimer, Phys. Rev. D **58**, 013009 (1998) [arXiv:astro-ph/9710115].
36. N. Iwamoto, Ann. Phys. **141**, 1 (1982)

NEW APPROACHES TO STRONG COUPLING LATTICE QCD

SHAILESH CHANRASEKHARAN
Department of Physics, Duke University, Durham, NC 27708-0305,
sch@phy.duke.edu

Received 1 December 2004

Efficient cluster algorithms have recently been discovered to solve strong coupling lattice QCD with staggered fermions in the chiral limit from first principles. This allows us for the first time to uncover the universal properties close to chiral phase transitions and make connections with chiral perturbation theory. In this article we will review some of the recent progress and outline some possible directions for future work.

Keywords: Lattice QCD; strong coupling; cluster algorithms; chiral limit.

1. Motivation

Non-perturbative calculations of Quantum Chromodynamics in the lattice formulation were first envisioned almost twenty five years ago.[1] Although there has been great progress in the field since then, it is still difficult to compute many quantities reliably due to errors from three sources: finite lattice spacings, finite volumes and unphysical quark masses. Today, these errors are reduced or eliminated by extrapolation techniques. Perturbative calculations allow us to extrapolate to the continuum limit, finite size scaling results help us to extrapolate to the infinite volume limit and predictions from chiral perturbation theory (χPT) are used for extrapolations to realistic quark masses. One of the contentious issues today is whether the quark masses that are currently used in the calculations are small enough for χPT to be applicable. Unfortunately, it is difficult to lower the quark masses without increasing the computational effort significantly. This problem is ubiquitous to all field theories which involve microscopic fermionic degrees of freedom; fermions make Monte Carlo algorithms notoriously difficult. The conventional approach is to integrate out the fermions in favor of a determinant which is used as a part of the Boltzmann weight during the process of a Monte Carlo update.[2,3] In many interesting situations, like in the case of QCD at finite baryon densities, the determinant is not even positive definite. This makes it impossible to design Monte Carlo algorithms and the problem is referred to as the *sign problem.* Fortunately there are examples where the determinant is positive and it is possible to design Monte Carlo algorithms. However, in these cases the number of computations necessary to compute the determinant during a single local Monte Carlo update increases with the system size. At small quark masses the determinant can also fluctuate a lot, which makes Monte Carlo algorithms difficult. Our inability to reliably compute quantities with sufficiently light quarks can be regarded as one of the outstanding problems in the field of lattice QCD.

Recently, it was discovered that the partition function of certain fermionic models can be rewritten as a sum over positive Boltzmann weights of configurations involving fermion world-lines.[4] Once a model is written in the world line representation it is often possible to find very efficient cluster algorithms to solve the problem.[5] Lattice QCD with staggered fermions in the strong coupling limit is such an example of a fermionic model which can be solved elegantly using these new ideas. Previously, strong coupling lattice QCD was solved using either mean field theory or with local Monte Carlo algorithms which became inefficient as the quark masses were lowered. Thus, the recent progress is significant; for the first time a model closely connected with lattice QCD can be solved elegantly close to the chiral limit. In this article we describe the recent progress. In section 2 we review strong coupling lattice QCD, in section 3 we discuss the problems with earlier algorithms and how the recently developed "directed loop algorithm" avoids these deficiencies. Section 4, contains a physics application; we show why chiral extrapolations can be tricky. In Section 5 outline some directions for the future.

2. Review of Strong Coupling Lattice QCD

The partition function of lattice QCD can be compactly written as

$$Z(T,m) = \int [dU][d\psi][d\overline{\psi}]\ \exp\Big(-\frac{1}{g^2} S_g[U] - S_f[\psi,\overline{\psi},U;T,m]\Big), \tag{1}$$

where S_g is the gauge field action, S_f is the fermion action and g is the bare gauge coupling. The physical continuum limit is obtained in the $g \to 0$ limit. On the other hand the strong coupling limit ($g \to \infty$), although unphysical, is much simpler since then we can ignore the effects of S_g. This is the limit where progress has been achieved when S_f is taken to be the staggered fermion action given by

$$S_f[\psi,\overline{\psi},U;T,m] = -\sum_{x,\mu} \frac{\eta_{x,\mu}}{2}\Big[\bar{\psi}_x U_{x,\mu}\psi_{x+\hat{\mu}} - \bar{\psi}_{x+\hat{\mu}} U^{\dagger}_{x,\mu}\psi_x\Big] - m\sum_x \bar{\psi}_x\psi_x, \tag{2}$$

where x refers to the lattice site on a periodic four-dimensional hyper-cubic lattice of size L along the three spatial directions and size L_t along the euclidean time direction. The index $\mu = 1,2,3,4$ refers to the four space-time directions, $U_{x,\mu}$ are $U(N)$ or $SU(N)$ group elements in the fundamental representation ($N \times N$ matrices) which represent gauge fields and $\psi_x, \bar{\psi}_x$ are the N-component staggered quark fields. We will assume that the gauge fields satisfy periodic boundary conditions while the quark fields satisfy either periodic or anti-periodic boundary conditions. The factors $\eta_{x,\mu}$ are the well-known staggered fermion phase factors which represent the Dirac matrices in a spin-diagonal representation. Using an asymmetry factor between space and time one can introduce a temperature in strong coupling QCD. After absorbing this factor into the definition of $\eta_{x,4}$ we choose $\eta^2_{x,\mu} = 1, \mu = 1,2,3$ (spatial directions) and $\eta^2_{x,4} = T$ (temporal direction), where the real parameter T is a coupling that controls the temperature. At fixed L_t, by varying T on asymmetric lattices ($L_t << L$) one can study finite temperature phase transitions.[6]

When the gauge group is $U(N)$ the partition function given in eq.(1) can be rewritten as a partition function of a monomer-dimer system given by[7]

$$Z(T,m) = \sum_{[n,b]} \prod_{x,\mu} (z_{x,\mu})^{b_{x,\mu}} \frac{(N-b_{x,\mu})!}{b_{x,\mu}!N!} \prod_x \frac{N!}{n_x!} m^{n_x}. \tag{3}$$

Here $n_x = 0,1,2,..,N$ refers to the number of monomers on the site x, $b_{x,\mu} = 0,1,2,...,N$ represents the number of dimers on the bond connecting x and $x+\hat{\mu}$, m is the monomer weight, $z_{x,\mu} = \eta^2_{x,\mu}/4$ are the dimer weights. Note that while spatial dimers carry a weight $1/4$, temporal dimers carry a weight $T/4$. The sum is over all monomer-dimer configurations $[n,b]$ which are constrained such that at each site, $n_x + \sum_\mu [b_{x,\mu} + b_{x-\hat{\mu},\mu}] = N$. Instead of $U(N)$ if the gauge group is chosen to be $SU(N)$, configurations can contain additional loops that form world-lines of baryons in addition to monomers and dimers. Since each baryon loop is made up of N quark, no dimers or monomers are allowed to touch a baryon loop. When N is even the baryons turn out to be bosons and it can be shown that all Boltzmann weights are positive definite. On the other hand for odd N baryons turn out to be fermions and baryonic loops can carry a negative sign. However, it is possible to identify every baryonic loop with a unique purely dimer loop. At zero baryon chemical potential this identification allows one to cancel all negative signs.[8] This strategy of identifying positive and negative configurations in order to cancel negative signs has since been used to solve a variety of sign problems.[9,10,4] Figure 1 shows two examples of $SU(3)$ configurations one with the baryonic loop and the other where the baryonic loop is replaced by a purely dimer loop.

Fig. 1. Example of $SU(3)$ configurations in the chiral limit: dimer-baryonic loop configuration (left) and a purely dimer configuration (right).

3. Directed Loop Algorithms

Strong coupling lattice QCD with staggered fermions were studied first in the mean field approximation.[11] Later, the dimer representation of eq.(3) was discovered and first principles calculations using Monte Carlo techniques were attempted.[7,8] The algorithms used involved only local updates in which a dimer was replaced by two monomers and versa. Clearly such an update would be forbidden at zero quark mass, which also implies that the algorithm would encounter severe critical slowing down as the mass is reduced. As far as we know the massless limit has remained unstudied. Clearly one needs a non-local update since in the massless limit the system is constrained. Figure 2 shows two possible local updates in $U(1)$ gauge theory. The update in which the two dimers are flipped is allowed in the massless theory. However, notice that this update will not be able to easily change the lower portion of the $U(1)$ configuration shown on the right side in figure 2.

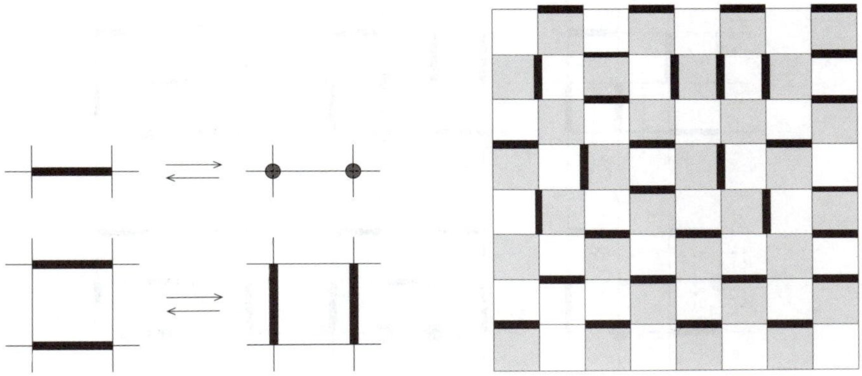

Fig. 2. Example of two local updates in the $U(1)$ theory (left) and an example of a $U(1)$ configuration in two dimensions (right). Note that in the massless theory the local updates shown on the left cannot update the lower portion of the configuration shown on the right.

Recently, it was discovered that a very efficient non-local update can be constructed for dimer systems.[12] This allows us for the first time to study the physics of the chiral limit of strong coupling lattice QCD. The algorithm is based on an idea that originated in quantum spin systems and involves sampling the partition function along with configurations that contribute to certain two point correlation functions.[13] For example consider the correlation function $G(x,y) = \langle\bar{\psi}(x)\psi(x)\ \bar{\psi}(y)\psi(y)\rangle$, where x and y are any two lattice sites. Configurations that contribute to $G(x,y)$ have monomers at x and y. At zero quark mass such configurations do not contribute to the partition function. This is the reason, it is very difficult to measure monomer-monomer correlations in the naive approach close to the chiral limit. In the new approach, called the directed loop algorithm, one would generate configurations that contain both zero monomers and two monomers. The algorithm can roughly be described as follows:[4]

- Pick a site at random and introduce a monomer on it. Introduce a second monomer on a neighboring site that is connected by a dimer and remove the dimer.
- While keeping the monomer on the first site as fixed move the other monomer around without allowing it to retrace the last step.
- The loop update ends when the moving monomer meets the original monomer so that both monomers are removed producing a new configuration without monomers.

An example of the directed loop update is shown in figure 3.

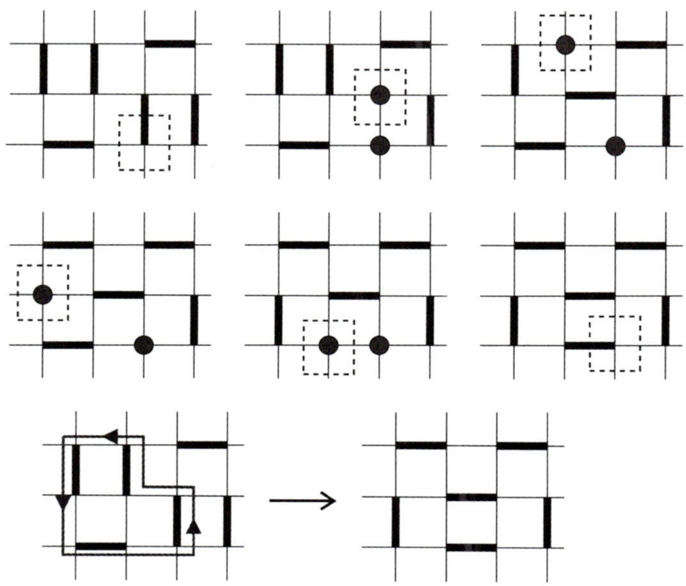

Fig. 3. Example of a directed loop update in the $U(1)$ theory.

Since detailed balance is satisfied at each step, a series of two monomer configurations that contribute to $G(x, y)$ have been sampled during the directed loop update. It is possible to argue that $\chi = \sum_x G(x, 0)$ is related to the size of the loop in the directed path algorithm.[12] Observables that depend on configurations that contribute to the partition function can be measured as usual. The directed loop algorithm works efficiently in the chiral limit and it can be modified to include a quark mass and can also be extended to any value of N.

4. Application: Reliability of Chiral Extrapolations

The directed loop algorithm has been recently applied to address a variety of physics questions related to the chiral limit of strong coupling Lattice QCD with $U(N)$ gauge fields. [14,15,16,17] Due to lack of space, here we will discuss only one application. As discussed in section 1, chiral perturbation theory helps us predict the form of chiral extrapolations in QCD. However, it is important to find the range of quark masses where these predictions can be useful and reliable. We can now address this question in strong coupling lattice QCD since we now have a very efficient algorithm with which we can study the chiral limit. We will see how a good algorithm allows one to get clear answers to such questions.

When $m = 0$ the staggered fermion action given in eq.(2) is invariant under $U(1)$ chiral transformations $\psi_x \rightarrow \exp(i\sigma_x\theta)\psi_x$ and $\bar{\psi}_x \rightarrow \bar{\psi}_x \exp(i\sigma_x\theta)$ where $\sigma_x = 1$ on all even sites and -1 on all odd sites. At low temperatures this chiral symmetry is spontaneously broken. In a recent study it was found that the symmetry is restored at a critical temperature T_c and the phase transition belongs to the 3dXY universality class as expected.[14] This means that at a fixed temperature below T_c, within the universal scaling window, the long distance physics of our model must be equivalent to a 3d $O(2)$ continuum field theory in its broken phase. The low energy effective action can be written in terms of $\vec{S}$, an $O(2)$ vector field with the constraint $\vec{S}\cdot\vec{S} = 1$. At the lowest order the effective action is given by [18]

$$S_{\text{eff}} = \int d^3x \Big[\frac{F^2}{2}\, \partial_\mu \vec{S}\cdot\partial_\mu\vec{S} + \Sigma\, \vec{h}\cdot\vec{S}\Big], \tag{4}$$

where $\vec{h}$ is the magnetic field and $|\vec{h}|$ is identified with the quark mass m. The two low energy constants F^2 and Σ that appear at this order are the pion decay constant and the chiral condensate respectively. Since we are discussing a three-dimensional effective theory, F^2 has dimensions of mass (or inverse length).

It is interesting to ask whether the results from the microscopic theory can be matched with the predictions of chiral perturbation theory. If computations were possible in the chiral limit ($m = 0$), then it is possible to find F^2 and Σ easily. In the microscopic theory one can define the helicity modulus

$$Y = \frac{1}{3L^3}\Big\langle \Big\{ [\sum_x J_{x,1}]^2 + [\sum_x J_{x,2}]^2 + [\sum_x J_{x,3}]^2 \Big\} \Big\rangle, \tag{5}$$

where $J_{x,\mu} = \sigma_x(b_{x,\mu} - N/8)$. When $m = 0$ the current $J_{x,\mu}$ is the conserved current associated with the $U(1)$ chiral symmetry. When $m = 0$ it is then possible to show that [18]

$$F^2 = \lim_{L\to\infty} Y. \tag{6}$$

One can also evaluate the chiral susceptibility,

$$\chi = \frac{1}{L^3}\frac{1}{Z}\frac{\partial^2}{\partial m^2} Z(T,m), \tag{7}$$

and use chiral perturbation theory to show[18]

$$\chi = \frac{\Sigma^2 L^3}{2}\Big[1 + \frac{0.226}{F^2 L} + \frac{\alpha}{L^2} + ...\Big], \tag{8}$$

where α depends on the higher order terms in the chiral Lagrangian. When $m \neq 0$ one can define

$$\Sigma_m = \lim_{L\to\infty} \frac{1}{L^3}\frac{1}{Z}\frac{\partial}{\partial m} Z(T,m). \tag{9}$$

Chiral perturbation theory then predicts that[18]

$$\Sigma_m = \Sigma\Big[1 + \frac{\sqrt{\Sigma}}{8\pi F^3}\sqrt{m} + b_2 m + b_3 m^{3/2} + ...\Big], \tag{10}$$

where b_2 and b_3 come from higher order terms in the chiral Lagrangian. Consistency of equations (6), (8), (10) can be used to check the reliability of chiral perturbation theory in a given range of quark masses.

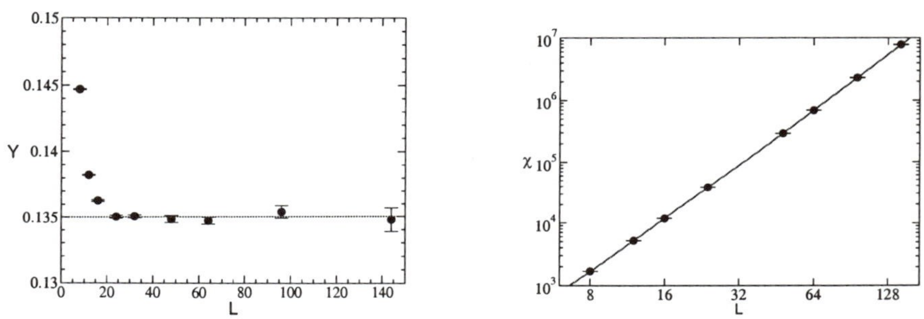

Fig. 4. Results for Y(left) and χ(right) as a function of the lattice size L. At large L we find $F^2 = Y = 0.1350(3)$. With F^2 fixed to 0.135, the data for χ fits very well to the form of eq.(8) and gives $\Sigma = 2.2648(10)$, $\alpha = 4.6(3)$ with a $\chi^2/DOF = 0.87$.

The calculations we discuss below were done on $L_t = 4$ lattices with $U(3)$ gauge fields. In this model it was found that $T_c = 7.47739(3)$.[14] In order to avoid very large correlation lengths while at the same time remaining in the 3dXY scaling window, we choose to work at $T = 7$. In figure 5 we plot our results of Y and χ for $m = 0$. We can use eq.(6) to compute F^2 and we find $F^2 = 0.1350(3)$. Using this value of F^2 we can fit the L dependence of our data for χ to the form given in eq.(8). The fit gives $\Sigma = 2.2648(10)$, $\alpha = 4.6(3)$ with a $\chi^2/DOF = 0.87$. Having established consistency at $m = 0$, we can turn to the quark mass dependence predicted in eq.(10). In figure 5 we plot our results. If we fix $\Sigma = 2.2648$ and $F^2 = 0.135$ we find that the fit of the data to the form of eq.(10) is good only when $m < 0.004$. We get $b_2 = 29.2(2)$ and $b_3 = -208(4)$ with a $\chi^2/DOF = 0.93$. Unfortunately, most lattice QCD calculations are done today with $m \geq 0.005$ in lattice units. A simple metropolis algorithm for strong coupling QCD also begins to fail below $m \sim 0.005$. Assuming that we are able to push these calculations up to $m = 0.001$, we can ask how the fit would look if we take into account only $m > 0.001$ data. Since Σ and F^2 would not be known, we choose them to be fit parameters. Again a very good fit is possible and we get $\Sigma = 2.218(3)$, $F^2 = 0.081(1)$, $b_2 = 2.5(5)$ and $b_3 = -18(2)$. This fit is shown as a dashed line in the figure. Clearly, although the fit is excellent it gives unreliable values for Σ and F^2.

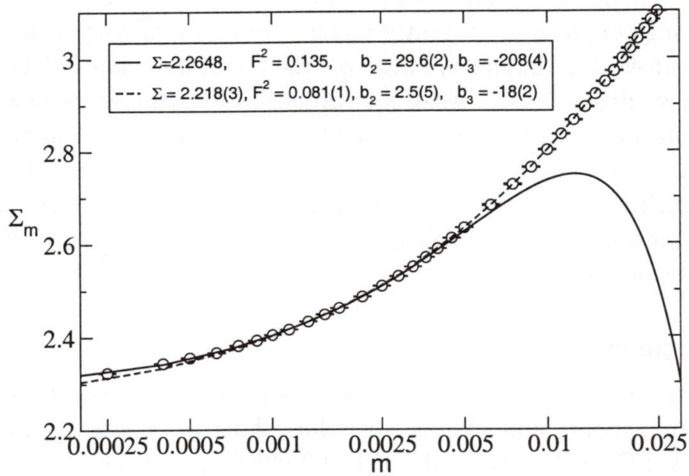

Fig. 5. The chiral condensate as a function of the quark mass. The solid line represents a fit to the small masses with $m < 0.004$ and $\Sigma = 2.2648$ and $F^2 = 0.135$ fixed. The dashed line represents a fit to masses $m > 0.001$ with floating values of Σ and F^2. The fit function is given in eq.(10).

The above results show that indeed matching with chiral perturbation theory is possible, however only for very small masses. The best matching is possible at $m = 0$. If the results at small masses are not known then the fitting can give parameters which contain large and uncontrolled systematic errors. This is important to keep in mind while extracting physical parameters by performing chiral extrapolations in lattice QCD at weaker couplings.

5. Directions for the Future

All the progress so far involves staggered fermions. An important question is whether it is possible to extend the new approach to Wilson fermions. It has already been shown that in two dimensions such an extension is indeed possible.[19] Extensions to four dimensions needs to be investigated.

The recent progress in solving strongly coupled lattice QCD with staggered fermions allows us to solve a variety of new problems. One such problem is two color QCD at strong couplings. Due to the fact that quarks and anti-quarks can be transformed into each other the global symmetry of two color QCD is $SU(2) \times U(1)$

and not $U(1)$.[20] One can also build theories with more interesting symmetry structure which can perhaps be solved using the new ideas. For example by working with N_f flavors of staggered fermions interacting with $U(1)$ gauge fields one can design models with $U(N_f) \times U(N_f)$ chiral symmetries. Conventional cluster algorithms are known to fail for such chiral theories.[21] The recent progress then suggests that in an appropriate formulation the directed loop algorithm may perhaps provide a new approach for constructing cluster algorithms for a whole class of chiral theories.

The phase diagram at finite temperatures and baryon densities in two color QCD at strong couplings is currently under investigation. Although these studies are possible at weaker couplings, until now the chiral limit has not been explored. Preliminary results indicate a weak first order finite temperature phase transition at zero chemical potential and a mean field transition at zero temperatures.

Today, a variety of fermionic theories are also solvable using similar new and powerful algorithms. However, most of the progress involves strongly coupled theories. An extension to weak coupling theories is desirable and success in this area will be highly valuable.

Acknowledgments

Various parts of the work reported here was done in collaboration with David Adams, Fu-Jiun Jiang and Costas Strouthos. This work was supported in part by grants DE-FG-96ER40945 and DE-FG02-03ER41241 from the Department of Energy (DOE).

References

1. M. Creutz, *Phys. Rev.* **D21**, 2308 (1980).
2. S. Duane, *Nucl. Phys.* **B257**, 652 (1985).
3. S. R. White, D. J. Scalapino, R. L. Sugar, E. Y. Loh, J. E. Gubernatis and R. T. Scalettar, *Phys. Rev.* **B40**, 506 (1989).
4. S. Chandrasekharan and U.-J. Wiese, *Phys. Rev. Lett.* **83**, 3116 (1999).
5. N. Prokof'ev and B. Svistunov, *Phys. Rev. Lett.* **87**, 160601 (2001).
6. G. Boyd, J. Fingberg, F. Karsch, L. Karkkainen and B. Petersson *Nucl. Phys.* **B376**, 199 (1992).
7. P. Rossi and U. Wolff, *Nucl. Phys.* **B248**, 105 (1984).
8. F. Karsch and H. Mütter, *Nucl. Phys.* **B313**, 541 (1989).
9. W. Bietenholz, A. Pochinsky and U. J. Wiese, *Phys. Rev. Lett.* **75**, 4524 (1995).
10. M. G. Alford, S. Chandrasekharan, J. Cox and U. J. Wiese, *Nucl. Phys. B* **602**, 61 (2001).
11. H. Kluberg-Stern, A. Morel, O. Napoly and B. Petersson, *Nucl. Phys.* **B190**, 504 (1981).
12. D.H. Adams and S. Chandrasekharan, *Nucl. Phys.* **B662**, 220 (2003).
13. O. F. Syljuasen and A. W. Sandvik, *Phys. Rev.* **E66**, 046701 (2002).
14. S. Chandrasekharan and F. J. Jiang, *Phys.Rev.* **D68**, 091501 (2003).
15. S. Chandrasekharan and C. G. Strouthos, *Phys. Rev.* **D68**, 091502 (2003).
16. S. Chandrasekharan and C. G. Strouthos, *Phys.Rev.* **D69**, 091502 (2004).
17. S. Chandrasekharan and C. G. Strouthos, arXiv:hep-lat/0410036.

18. P. Hasenfratz and H. Leutwyler, *Nucl. Phys.* **B343**, 241 (1990).
19. M. Salmhofer, *Nucl. Phys.* **B362**, 641 (1991).
20. S. Hands, J. B. Kogut, M. P. Lombardo and S. E. Morrison, *Nucl.Phys.* **B558**, 327 (1999).
21. S. Caracciolo, R. G. Edwards, A. Pelissetto and A. D. Sokal, *Nucl. Phys.* **B403**, 475 (1993).

NUCLEAR INTERACTIONS FROM THE RENORMALIZATION GROUP

ACHIM SCHWENK*

Department of Physics, The Ohio State University, Columbus, OH 43210

Received 1 December 2004

We discuss how the renormalization group can be used to derive effective nuclear interactions. Starting from the model-independent low-momentum interaction $V_{\text{low}\,k}$, we successively integrate out high-lying particle and hole states from momentum shells around the Fermi surface as proposed by Shankar. The renormalization group approach allows for a systematic calculation of induced interactions and yields similar contributions to the scattering amplitude as the two-body parquet equations. We review results for the $^1S_0$ and $^3P_2$ superfluid pairing gaps as well as the spin dependence of effective interactions in neutron matter. Implications for the cooling of neutron stars are discussed.

1. Introduction

Conventional precision nucleon-nucleon (NN) interactions are well-constrained by two-nucleon scattering data only for laboratory energies $E_{\text{lab}} \lesssim 350\,\text{MeV}$. As a consequence, details of nuclear forces are not constrained for relative momenta $k > 2.0\,\text{fm}^{-1}$ or for relative distances $r < 0.5\,\text{fm}$. All these NN potentials have repulsive cores, which lead to significant probing of the model-dependent high-momentum components in few and many-body applications. This model dependence, when not compensated by three-body interactions, results in the Tjon and Coester lines.

Using the renormalization group (RG), we have integrated out the high-momentum modes above a cutoff Λ in momentum space. The resulting low-momentum interaction, called $V_{\text{low}\,k}$, only has momentum components below the cutoff and evolves with Λ so that all low-energy two-body observables below the cutoff (phase shifts and deuteron binding energy) are invariant. We have shown that for $\Lambda \lesssim 2.0\,\text{fm}^{-1}$, all NN potentials that fit the scattering data and include the same long-distance pion physics lead to the same "universal" low-momentum interaction $V_{\text{low}\,k}$.[1,2] When $V_{\text{low}\,k}$ is augmented by a low-momentum three-nucleon (3N) force, which regulates $A = 3, 4$ binding energies, we find that the 3N parts are perturbative for $\Lambda \lesssim 2.0\,\text{fm}^{-1}$.[3] By perturbative we mean $\langle \Psi^{(3)} | V_{3N} | \Psi^{(3)} \rangle \approx \langle \Psi^{(2)} | V_{3N} | \Psi^{(2)} \rangle$, where $|\Psi^{(n)}\rangle$ are exact solutions including up to n-body forces.

Since $V_{\text{low}\,k}$ does not have a strong core at short distances, it can be used directly in many-body applications without a G matrix resummation. Fig. 1 shows the

*Present address: Nuclear Theory Center, Indiana University, Bloomington, IN 47408.

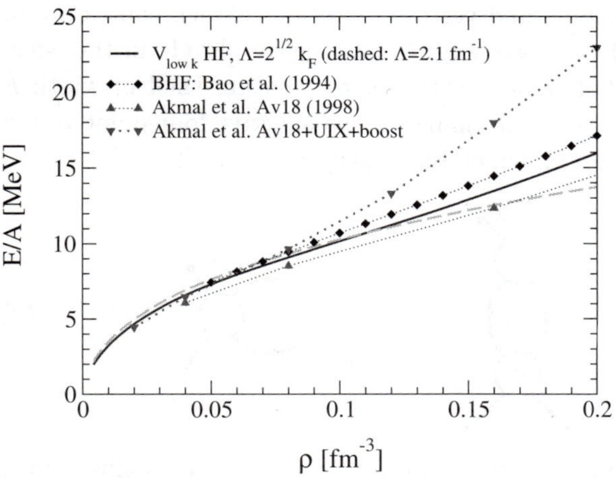

Fig. 1. Comparison of the $V_{\text{low}\,k}$ HF EoS for pure neutron matter with Brueckner-Hartree-Fock (BHF) results using the Bonn potential and results of Akmal *et al.* obtained using chain summation methods, for references see.[4] Calculations indicate that nuclear matter may be perturbative with low-momentum interactions.[5] Contrary to all other microscopic interactions, $V_{\text{low}\,k}$ yields bound nuclei on the HF level.[6]

equation of state (EoS) of pure neutron matter obtained in the Hartree-Fock (HF) approximation.[4] In symmetric nuclear matter, saturation is due to the 3N force.[5]

2. Renormalization Group Approach to Neutron Matter

A central challenge in nuclear many-body theory lies in understanding nuclear data and predicting input for astrophysical simulations from microscopic nuclear interactions. In these proceedings, we review results for the superfluid pairing gap, where many-body correlations are crucial even in the perturbative limit, due to the singular dependence of the gap on the interaction. For dilute neutrons interacting through an attractive S-wave scattering length $a_{\rm S} < 0$ only, the pairing gap is given by[7]

$$\Delta = \frac{8}{e^2}\,\varepsilon_{\rm F} \exp\Big\{ \text{const.} \Big($$

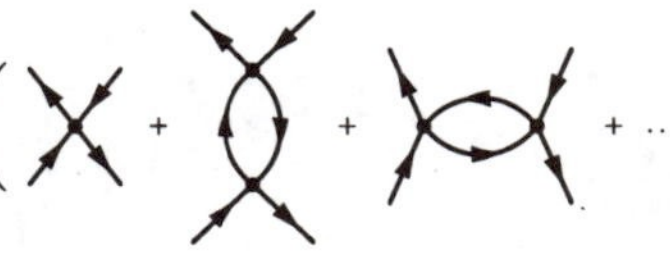

$$\Big)^{-1} \Big\}$$

$$= \frac{8}{e^2}\,\varepsilon_{\rm F} \exp\bigg(\frac{2}{\pi k_{\rm F} a_{\rm S}} + \log(0.45) + \mathcal{O}(k_{\rm F} a_{\rm S}) \bigg), \qquad (1)$$

where the second-order diagrams are projected on S-wave and e is the exponential constant. Thus, particle-hole (ph) polarization effects reduce the pairing gap by a factor $(4e)^{-1/3} \approx 0.45$ compared to the mean-field BCS estimate, due to long-range spin fluctuations. In neutron matter, it is therefore crucial to include ph correlations through the induced interaction for a realistic assessment of polarization effects.

In many-body systems, the RG provides a systematic tool to construct non-perturbative effective interactions among valence nucleons. We have applied the RG to neutron matter, restricting the effective interaction to low-lying states in the vicinity of the Fermi surface.[4] This follows the RG approach to interacting Fermi systems proposed by Shankar.[8] The method is widely used in condensed matter physics to study the interference of different instabilities, especially in the context of the 2d Hubbard model.[9] Starting from $V_{\text{low}\,k}$ in the full space, the RG generates the induced interaction, i.e., screening and vertex corrections, which contribute to the quasiparticle interaction and the low-energy scattering amplitude in the effective theory defined for particle/hole modes within a momentum shell of width Λ around the Fermi surface. At one loop, the change of the effective four-point vertex $a(\mathbf{q}, \mathbf{q}'; \Lambda)$ is given by an RG equation, which reads diagrammatically

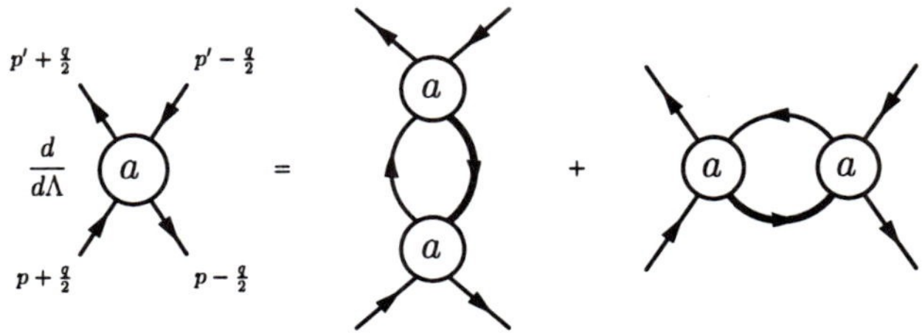

(2)

Here, the thin lines denote intermediate states from thin momentum shells, which are integrated out successively, and the thick lines denotes "fast" particles/holes with $|p_i - k_\mathrm{F}| \geqslant \Lambda$. The RG treats the dependence on the momenta $\mathbf{q}$ and $\mathbf{q}' = \mathbf{p} - \mathbf{p}'$ on an equal footing and maintains the symmetries of the scattering amplitude. Currently, we treat pairing correlations explicitly in weak-coupling BCS theory, after the RG is evolved to the Fermi surface. Future directions will include the

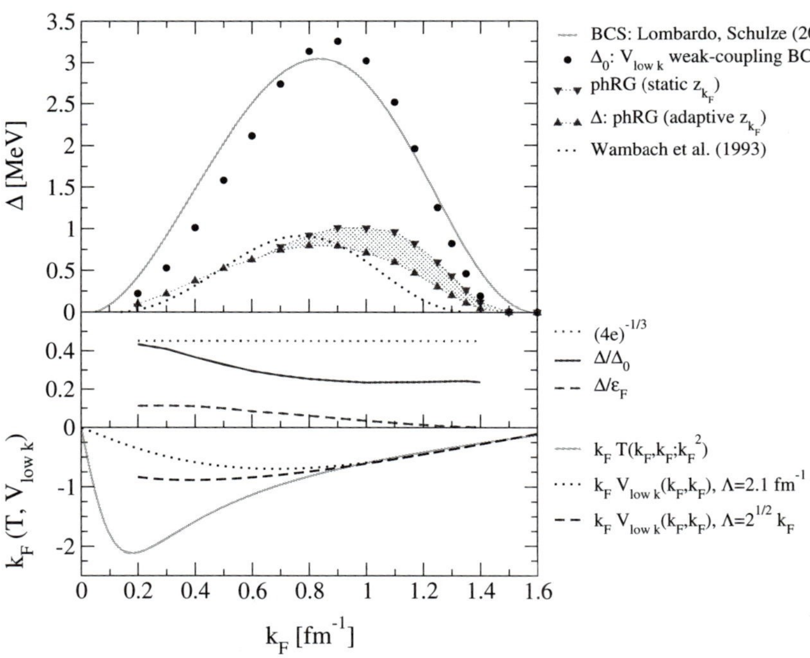

Fig. 2. Upper panel: Comparison of mean-field BCS results for the $^1\mathrm{S}_0$ superfluid gap to results including polarization effects through the phRG, for references and further details, including the z_{k_F} factor, see.[4] Middle panel: Comparison of the full superfluid gap to the mean-field BCS gap and the Fermi energy. We also show the universal extreme low-density limit, $\Delta/\Delta_0 = (4e)^{-1/3} \approx 0.45$. Lower panel: Comparison of dimensionless lowest-order pairing interactions.

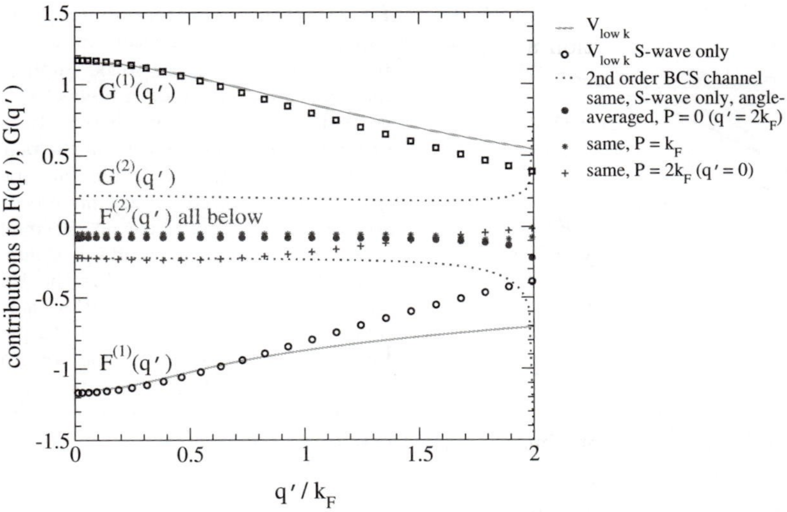

Fig. 3. Comparison of $V_{\mathrm{low}\,k}$ (1) and second-order BCS channel contributions (2) to the scalar, $F(q')$, and spin-spin dependent Landau functions, $G(q')$, in neutron matter for $k_{\mathrm{F}} = 1.0\,\mathrm{fm}^{-1}$. The lines are from[13] and the points are S-wave approximations thereof. The differences on the $V_{\mathrm{low}\,k}$ level are due to contributions from higher partial waves. Small $q' \to 0$, where higher partial waves are negligible in $F^{(1)}(q')$, correspond to $P \to 2k_{\mathrm{F}}$. Differences for $F^{(2)}(q')$ for larger q' are therefore due to higher partial waves, angle-averaging, as well as hole-hole contribions, which are included in the dotted lines, but not the points, and are also absent for $P \to 2k_{\mathrm{F}}$. In contrast to free-space scattering, where the second Born term is comparable to $V_{\mathrm{low}\,k}$, we find that the second-order particle-particle contributions are small, except for low-lying pairing correlations ($\sim \log(P/k_{\mathrm{F}})$).

BCS channel in the RG equation. On the Fermi surface, $\mathbf{q}$, $\mathbf{q}'$ and $\mathbf{P} = \mathbf{p} + \mathbf{p}'$ are orthogonal, and therefore, we approximate $a(\mathbf{q}, \mathbf{q}'; \Lambda) = a(q^2, q'^2; \Lambda)$ in the RG to extrapolate off the Fermi surface. On the $V_{\mathrm{low}\,k}$ level, the $\mathbf{q} \cdot \mathbf{q}'$ dependence is small.

The efficacy of the RG lies in including many-body correlations from successive momentum shells, on top of an effective interaction with particle/hole polarization effects from all previous shells. By solving Eq. (2) iteratively, it can be seen that the RG builds up correlations similar to the two-body parquet equations, see also Ref. 10.

In neutron matter, non-central and 3N forces are weaker, and we have solved the RG equations with the initial condition $a(\Lambda = k_{\mathrm{F}}) = V_{\mathrm{low}\,k}$ including only scalar and spin-spin interactions, which dominate at low densities. In addition to the Fermi liquid parameters (see Fig. 6 in Ref. 4), the RG solution yields the scattering amplitude for finite scattering angles, which is needed, e.g., for transport and pairing. In Fig. 2, we present our results for the ${}^1\mathrm{S}_0$ pairing gap. We find a suppression of the S-wave gap due to spin fluctuations from $\Delta_0 \approx 3.3\,\mathrm{MeV}$ to $\Delta \approx 0.8\,\mathrm{MeV}$ at maximum. Our results are similar to those of Wambach *et al.*,[11] and our mean-field ($V_{\mathrm{low}\,k}$) weak-coupling gap agrees well with the BCS result. This is consistent with the agreement of the $V_{\mathrm{low}\,k}$ HF EoS with the BHF results and shows that $V_{\mathrm{low}\,k}$ does not lead to strong short-range correlations. In Fig. 3, we also compare the S-wave second-order particle-particle contributions to the full second-order result.

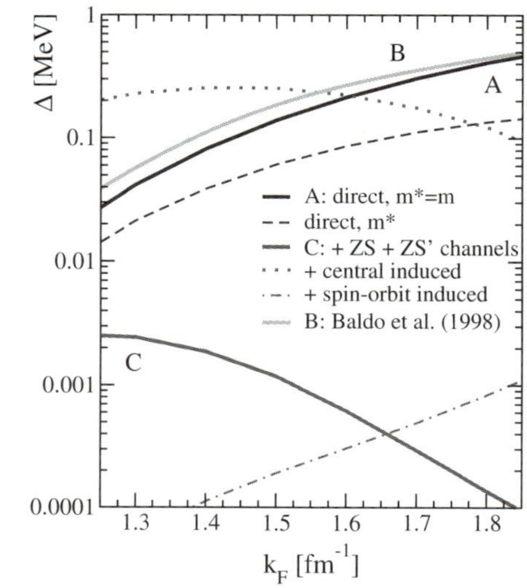

Fig. 4. The angle-averaged $^3P_2$ superfluid gap versus Fermi momentum in neutron matter. The direct ($V_{\text{low}\,k}$) weak-coupling result for a free spectrum and with effective mass is compared to the gap which includes second-order ph polarization effects on the pairing interaction. We also show the modification of the gap, if only induced central or only induced spin-orbit effects are taken into account. As a reference, we give the results of Baldo *et al.*, which are obtained by solving the BCS gap equation in the coupled $^3P_2$–$^3F_2$ channel for different free-space interactions. As predicted by Pethick and Ravenhall, and confirmed by our results, superfluidity is enhanced, if only central induced interactions were included. For details and references see.[13]

3. Spin-dependence of Effective Interactions

Non-central forces in nuclear matter lead to novel spin-dependencies due to the presence of the Fermi sea. In addition to the standard forces, novel spin non-conserving effective interactions, $i(\boldsymbol{\sigma}_1 - \boldsymbol{\sigma}_2) \cdot \mathbf{q} \times \mathbf{P}$ and $(\boldsymbol{\sigma}_1 \times \boldsymbol{\sigma}_2) \cdot (\mathbf{q}' \times \mathbf{P})$, are induced by screening in the ph channels and a novel long-wavelength tensor force $S_{12}(\mathbf{P})$ is generated.[13] Moreover, due to the coupling of the tensor and spin-orbit force to the strong spin-spin interaction G_0, we find that the tensor component of the quasiparticle interaction and the P-wave spin-orbit pairing force are significantly reduced in neutron matter, see Figs. 1 and 2 in Ref. 13 As illustrated in Fig. 4, the screening of the spin-orbit interaction leads to a strong suppression of the $^3P_2$ superfluid gap for neutrons in the interior of neutron stars. Note that in vacuum the spin-orbit force is crucial for a realistic description of the P-wave phase shifts and that the suppression of the gap is due to a reduction of the pairing interaction by only $< 50\%$.

Typical interior temperatures of isolated neutron star sources are on the order of $T \sim 10^8\,\text{K} \approx 10\,\text{keV}$ and therefore it is possible to constrain the P-wave gaps phenomenologically through neutron star cooling simulations. It has been found that with free-space gaps $\Delta_0 \sim 0.1\,\text{MeV}$, neutron stars cool too rapidly and that a consistency with the data requires $^3P_2$ gaps $\Delta \lesssim 30\,\text{keV}$ [14] or $\Delta \lesssim 10\,\text{keV}$.[15] A less severe but still relevant dependence on the gap has been found in Ref. 16 The same microscopic scattering amplitudes can be used to constrain neutrino emissivities.[17] Finally, polarization effects and novel spin-dependences are important for magnetic susceptibilities and long-wavelength spin-isospin response, as discussed in Ref. 18.

Work is in progress to extend the RG approach to non-central interactions and to asymmetric matter including neutron/proton pairing.[12] In 2d, higher loops in the RG equation are geometrically suppressed by Λ/k_F for regular Fermi surfaces.[8] This may enable an error estimate of the one-loop truncation for low-momentum interactions, which do not scatter strongly to high-lying (high Λ) states. Finally, the RG approach allows for a consistent renormalization of currents/operators and an extension to effective valence shell-model interactions for heavy nuclei.

I am grateful to Scott Bogner, Gerry Brown, Bengt Friman, Dick Furnstahl, Chuck Horowitz, Tom Kuo, Emma Olsson, Janos Polonyi and Chris Pethick for many useful discussions. This work was supported by the NSF under grant No. PHY-0098645 .

References

1. S.K. Bogner, T.T.S. Kuo, A. Schwenk, D.R. Entem and R. Machleidt, *Phys. Lett.* **B576**, 265 (2003), for details on the RG see nucl-th/0111042.
2. S.K. Bogner, T.T.S. Kuo and A. Schwenk, *Phys. Rep.* **386**, 1 (2003).
3. A. Nogga, S.K. Bogner and A. Schwenk, *Phys. Rev.* **C** (R) in press, nucl-th/0405016.
4. A. Schwenk, B. Friman and G.E. Brown, *Nucl. Phys.* **A713**, 191 (2003), for a short discussion see also nucl-th/0302081, in Hirschegg 2003.
5. S.K. Bogner, A. Schwenk, R.J. Furnstahl and A. Nogga, in prep.
6. L. Coraggio *et al.*, *Phys. Rev.* **C68**, 034320 (2003) and nucl-th/0407003.
7. L.P. Gorkov and T.K. Melik-Barkhudarov, *Sov. Phys. JETP* **13**, 1018 (1961), see also H. Heiselberg, C.J. Pethick, H. Smith and L. Viverit, *Phys. Rev. Lett.* **85**, 2418 (2000).
8. R. Shankar, *Rev. Mod. Phys.* **66**, 129 (1994).
9. C. Honerkamp, these proceedings, cond-mat/0411267, and references therein.
10. A. Schwenk, G.E. Brown and B. Friman, *Nucl. Phys.* **A703**, 745 (2002).
11. J. Wambach, T.L. Ainsworth and D. Pines, *Nucl. Phys.* **A555**, 128 (1993).
12. A. Schwenk, B. Friman and R.J. Furnstahl, in prep.
13. A. Schwenk and B. Friman, *Phys. Rev. Lett.* **92**, 082501 (2004).
14. D.G. Yakovlev and C.J. Pethick, *Ann. Rev. Astron. Astrophys.* **42**, 169 (2004).
15. D. Blaschke, H. Grigorian and D.N. Voskresensky, *Astron. Astrophys.* **424**, 979 (2004).
16. D. Page, J.M. Lattimer, M. Prakash and A.W. Steiner, astro-ph/0403657.
17. A. Schwenk, P. Jaikumar and C. Gale, *Phys. Lett.* **B584**, 241 (2004).
18. E. Olsson and C.J. Pethick, *Phys. Rev.* **C66**, 065803 (2002), E. Olsson, P. Haensel and C.J. Pethick, *Phys. Rev.* **C70**, 025804 (2004) and private communication.

RANDOM INTERACTIONS AND GROUND STATE SPIN OF FINITE FERMI SYSTEMS

VLADIMIR ZELEVINSKY

National Superconducting Cyclotron Laboratory and Department of Physics and Astronomy, Michigan State University, East Lansing, MI 48824-1321, USA
zelevinsky@nscl.msu.edu

ALEXANDER VOLYA

Department of Physics, Florida State University, Tallahassee, FL 32306-4350, USA
volya@martech.fsu.edu

Received 1 December 2004

Numerical simulations show that the ground state spin of a finite many-body system with random but rotationally invariant interactions is predominantly zero. This means that it is not sufficient to explain the zero ground state spin of even-even nuclei only by pairing correlations. Simple explanations based on a statistical multiplicity, width of the level density, effective pairing, formation of condensates etc. do not work. We suggest statistical theory based on the effective mean field and geometric chaoticity of angular momentum coupling of individual particles. The results show a significant role of symmetry arguments in finite systems.

Keywords: Random interactions; ground state spin; chaos; many-body system.

1. The Problem

The problem we discuss below emerges on the crossing of two subfields of quantum physics, many-body theory and quantum chaos. We consider a finite system of interacting constituents – a complex atom, molecule, nucleus, atomic cluster, or quantum dot. As excitation energy and level density increase, all interactions become effectively strong. The many-body stationary states are gradually transformed into exceedingly complicated superpositions of a large number of simple basis configurations, for example Slater determinants in the mean field basis for Fermi systems. This is a region of *many-body quantum chaos.*[1] The spectral statistics and average properties of the wave functions here are essentially independent of the details of interactions being close to the extreme limit of random matrix theory.[2] It is usually assumed that at this stage the constants of motion, except for energy, are destroyed by chaotic mixing. In practice in finite systems total spin **J** is strictly conserved by any rotationally invariant interaction, and the states are divided into classes with definite values of J and J_z (plus more specific conserved quantities as isospin in nuclei). This does not matter for chaos that takes place independently in each class of states.

Now we can look at the situation near the *ground state*. Its structure is usually thought to be defined by the most coherent parts of the interaction. The spin interaction decides for the system to be ferro- or antiferro-magnetic, while the thermal excitation destroys magnetic order. An empirical fact that all even-even nuclei have ground state spin $J_0 = 0$ is traditionally considered as a result of the pairing correlations of superconducting type, attraction in a particle-particle channel with angular momentum $L = 0$ and isospin $T = 1$ of the pair. By this logic, eliminating pairing one could get a different value of J_0. It is known, for example, that in the models of quantum spin glasses with random sign of spin-spin interaction,[3] the ground state spin can take various values growing in average $\propto \sqrt{N}$ with the number N of interacting spins.

In the absence of coherent, for instance pairing-like, interactions, one would expect a similar random-walk result in other cases. This however turns out to be wrong as first was shown in Ref. 4. Instead, the ground state spin in the random ensemble of *rotationally invariant* interactions is predominantly zero. We come here to a problem never discussed before in relation to quantum chaos. Various sectors of Hilbert space are governed by the same Hamiltonian. Therefore their spectra are *correlated* even in the chaotic limit. Below we show the results of numerical simulations and discuss the underlying physical picture. The details and many references can be found in recent review papers [5] and [6] written from rather different positions.

2. Observations

Let us consider the simplest case: N fermions are put on a single orbital with a given value of the total (half-integer) angular momentum j. The single-particle states $|jm)$ are degenerate. The two-body rotationally invariant interaction can be presented as a sum of the channels with spin L and its projection Λ of the pair,

$$H = \sum_{L\Lambda} V_L P^\dagger_{L\Lambda} P_{L\Lambda}, \tag{1}$$

where the normalized pair operators are defined with the use of standard vector coupling, $P^\dagger_{L\Lambda} = (1/\sqrt{2})[a^\dagger_j a^\dagger_j]_{L\Lambda}$. Due to Fermi statistics, only even values $L = 0, 2, ..., 2j-1$ are allowed, and the amplitudes V_L are treated as uncorrelated random parameters. The results are not strongly sensitive to their distributions, and we can use either Gaussian or uniform, from -1 to +1, distributions always keeping $\langle V_L \rangle = 0$.

Typical results are presented in Tab. 1 for 6 particles at the $j = 21/2$ orbit. If all ground state spins would appear randomly, we would expect their frequency to agree with the statistical multiplicity $d(J)$ in Hilbert space (the maximum at $J = 16$ in the example shown in Tab. 1, see column (i)). The numerical experiment clearly shows the predominance of $J = 0$; the probability of the maximum allowed spin is also enhanced. The obvious idea of the role of pairing does not work: all cases of excluded pairing, maximum attraction and maximum repulsion in the $L = 0$ channel display qualitatively similar picture, although the exact value of f_0 changes